# GRAPHS AND TABLES OF THE
# MATHIEU FUNCTIONS
# AND THEIR FIRST DERIVATIVES

BY

JAMES C. WILTSE

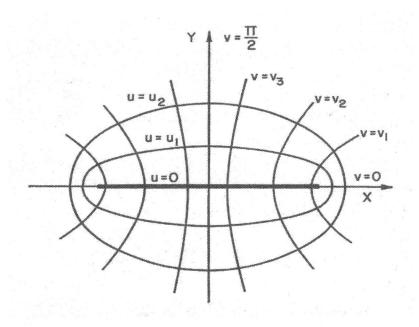

*AuthorHouse™*
*1663 Liberty Drive*
*Bloomington, IN 47403*
*www.authorhouse.com*
*Phone: 1-800-839-8640*

*First published by AuthorHouse    01/16/2012*

*ISBN: 978-1-4685-4462-6 (sc)*
*ISBN: 978-1-4685-4461-9 (ebk)*

*Library of Congress Control Number: 2012900919*

*Printed in the United States of America*

# Graphs and Tables of the
# Mathieu Functions
# and their First Derivatives

**James C. Wiltse**

authorHOUSE®

## Acknowledgement

Dr. Marcia J. King made significant contributions to this document. She calculated many of the tabulated values and verified the analysis for Section I and II.  Her assistance was much appreciated.

The author would like to acknowledge the editorial assistance of Dr. Helen C. Wiltse in preparing the material for this volume.

# TABLE OF CONTENTS

## Foreword

Mathieu functions are employed in solving a variety of problems in mathematical physics. In many cases the configuration involves elliptical coordinates. Of course, the circular geometry is the degenerate case of the elliptical cross section.

A typical example is that of cylindrical waveguides for electromagnetic waves. This includes hollow metal guides, coaxial transmission lines, and surface waveguides (a dielectric rod or a metal wire) of elliptical cross section. Solutions of the wave equation involve products of angular periodic and radial Mathieu functions. Similarly, circular waveguides involve products of sinusoids and Bessel functions, which the Mathieu functions smoothly transition to as the ellipse becomes a circle.

This volume contains values for, and curves of the angular and radial Mathieu functions and their first derivatives. The latter are often needed in the solution of problems, in particular in solving electromagnetic wave propagation problems. Also included are data on zero crossings of the radial Mathieu functions. These are often needed for determining the cut-off frequencies for propagating modes.

Other tables are available for the Mathieu functions, but there is very little data available for derivatives or zero crossings. It is felt that the principal value of this volume is in the multitude of curves included. The analyst dealing with elliptical cases can, by inspection of the curves, find values of the functions and derivatives at the origin, maxima and minima, zero crossings, and qualitative behavior of the plots as a function of several parameters. To the author's knowledge, this is the most extensive presentation of plotted information. It is hoped that the information will be helpful in the solution of practical problems.

This book is divided into two sections. Section I deals only with the functions themselves, defining the equations and terminology used and presenting the tabular data and curves. Section II treats the derivatives and the zeros. Again the terminology and equations for the first derivatives are given.

The Mathieu functions are named after Emile L. Mathieu (1835-1890), a French mathematician, who in 1868 published an article dealing with vibratory movement of the elliptic membrane. The asteroid 27947 Emilemathieu is named in his honor.

# Section I
## Periodic Angular and Radial Mathieu Functions

## Summary

This section contains tabulated values and curves for some of the Mathieu functions. Included are the periodic Mathieu functions $Se_n(s, v)$ and $So_n(s, v)$ for integer orders $0 \le n \le 2$ and various positive and negative real values of the parameter s. Also. included are the radial Mathieu functions $Je_n(s,u)$, $Jo_n(s, u)$, $Ne_n(s,u)$, and $No_n(s, u)$ with s positive real, and $He_n^{(1)}(s,u)$ and $Ho_n^{(1)}(s,u)$ with s negative real, for $0 \le n$ (integer) $\le 2$ and $0 \le u \le 2$ in each case.

# LIST Of
# ILLUSTRATIONS

## INTRODUCTION

The Mathieu functions appear in a large variety of physical problems[1,2,3], but in spite of this there is a paucity of tabulated values. The present report contains tables and curves of those periodic and radial functions which are particularly applicable to certain wave propagation problems. In the literature some variation exists in the notations and the normalization used for defining the various types of Mathieu functions. The definitions used herein are the same as the ones given in Reference 4 and are consistent with those of other American authors[3,5,6].

The two dimensional scalar wave equation

$$\nabla^2 \psi + k^2 \psi = 0 \ , \quad \psi = \psi(u,v) \tag{1}$$

when expressed in terms of the coordinates of an elliptic cylinder system such as that shown in Figure 1, becomes

$$\frac{\partial^2 \psi}{\partial u^2} + \frac{\partial^2 \psi}{\partial v^2} + k^2 d_o^2 \ (\cosh^2 u - \cos^2 v) \ \psi = 0 \tag{2}$$

where   $d_o$ = the semi-focal length of the ellipse,

   $u$ = the radial variable of the coordinate system,

   $v$ = the angular variable.

By letting $\psi(u,v) = f(u) \cdot g(v)$ and substituting this into Equation (2) the variables may be separated, and we obtain the following equations:

$$\frac{1}{f} \frac{\partial^2 f}{\partial u^2} + k^2 d_o^2 \cosh^2 u = \frac{-1}{g} \frac{\partial^2 g}{\partial v^2} + k^2 d_o^2 \cos^2 v = b \tag{3}$$

9

$$\frac{\partial^2 g}{\partial v^2} + (b - s \cos^2 v) \, g = 0 \tag{4}$$

$$\frac{\partial^2 f}{\partial u^2} - (b - s \cosh^2 u) \, f = 0 \tag{5}$$

where b is the separation constant or characteristic value and $s = k^2 d_o^2$.
Only real values of s and b have been used in the present report.
For a discussion of the case when s is imaginary, see Reference 1,
pages 48 and 320.

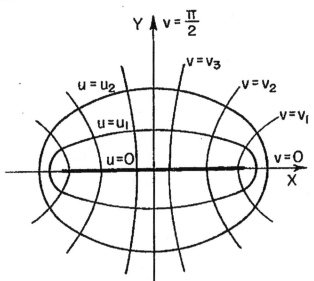

FIG. 1   THE ELLIPTIC CYLINDER COORDINATE SYSTEM.   THE
POSITIVE z DIRECTION IS OUT OF THE PAGE.

Equation (4) is a form of Mathieu's equation, while Equation
(5) is referred to as the modified Mathieu equation.   The angular
solutions which satisfy Equation (4) and which also have a period $\pi$ or
$2\pi$ are commonly called the periodic Mathieu functions and are

designated by $Se_n$ or $So_n$, depending upon whether the solutions are even or odd, respectively, in v. The periodic functions are represented by the following Fourier series:

$$Se_{2r}(s, v) = \sum_{k=0}^{\infty} De_{2k}^{(2r)} \cdot \cos 2kv \qquad \text{(period } \pi\text{)} \qquad (6)$$

$$Se_{2r+1}(s, v) = \sum_{k=0}^{\infty} De_{2k+1}^{(2r+1)} \cdot \cos (2k+1)v \qquad \text{(period } 2\pi\text{)} \qquad (7)$$

$$So_{2r}(s, v) = \sum_{k=1}^{\infty} Do_{2k}^{(2r)} \cdot \sin 2kv \qquad \text{(period } \pi\text{)} \qquad (8)$$

$$So_{2r+1}(s, v) = \sum_{k=0}^{\infty} Do_{2k+1}^{(2r+1)} \cdot \sin (2k+1)v \qquad \text{(period } 2\pi\text{)} \qquad (9)$$

The coefficients $De_m^{(n)}$ and $Do_m^{(n)}$ are dependent upon s and n, the order of the appropriate periodic Mathieu function. In practice the superscripts which indicate the order are often omitted from the coefficients. Reference 4 contains tables of values of $De_m$, $Do_m$, and the characteristic values , b, used in Equations (3), (4), and (5).

As $d_o$ and s go to zero, the ellipse becomes a circle and the periodic functions reduce to the trigonometric functions:

$$Se_n(s \to 0, v) \to \quad \cos nv \qquad (10)$$

$$So_n(s \to 0, v) \to \quad \sin nv \qquad (11)$$

The $De_m$ and $Do_m$ coefficients have been so normalized[*] that $Se_n(s, v)$

_____

* - See Morse and Feshbach (Reference 3), pages 1409 and 1411, for a discussion of the reasons for this choice of normalization.

and the derivatives of $So_n(s,v)$ with respect to $v$ are each unity at $v=0$, whether $s$ is zero or not.

Table I and Figures 2 through 15 give data on $Se_n(s,v)$ and $So_n(s,v)$ for $0 \leq n \leq 2$ and for various positive and negative real values of $s$. For many cases the angles were so chosen that, in conjunction with the table on pages 1934 and 1935 of Reference 3, one may obtain $Se_n(s,v)$ and $So_n(s,v)$ at 5-degree intervals in the angular variable $v$.

The solutions of Equation (5), the modified Mathieu equation, are known as the radial Mathieu functions. These have various analogies with, and may be expressed in terms of, the Bessel functions. The radial Mathieu functions of the first kind were calculated from the following expressions:

$$Je_{2r+p}(s,u) = (-1)^r \sqrt{\frac{\pi}{2}} \sum_{k=o}^{\infty} (-1)^k De_{2k+p} \cdot J_{2k+p}(\sqrt{s}\cosh u) \qquad (12)$$

$$Jo_{2r+p}(s,u) = (-1)^r \sqrt{\frac{\pi}{2}} \tanh u \sum_{k=o}^{\infty} (-1)^k (2k+p) \cdot Do_{2k+p} \cdot J_{2k+p}(\sqrt{s}\cosh u) \qquad (13)$$

where $J_{2k+p}(\sqrt{s}\cosh u)$ is the Bessel function of the first kind and of order $2k+p$, $p=0$ or $1$, and the $De_m$ and $Do_m$ are the coefficients used in calculating the periodic Mathieu functions of corresponding order. Table II and Figures 16 through 25 give data on $Je_n(s,u)$ and $Jo_n(s,u)$ for $0 \leq n \leq 2$, $0 \leq u \leq 2$, and various positive values of $s$. Note that the term "even (or "odd") as applied to any of the radial functions indicates that the radial solution was obtained from Equation (5) for the same characteristic value b which would give an even (or odd) periodic solution for Equation (4). For example, given a value of $s = s_a$, there is a number $be_n$ (where n is an assigned integer) which is a particular value of b, and for which $Se_n(s_a,v)$ is a solution of the Mathieu equation. If the same value of b is used in the modified Mathieu equation, the radial solution of the first kind will then be

$Je_n(s_a, u)$. Also, both $Je_n$ and $Se_n$ (or $Jo_n$ and $So_n$) are constructed from the same $De_m$ (or $Do_m$) coefficients.

A second, independent solution of the modified Mathieu equation also is of use in physical problems. The radial Mathieu functions of the second kind are given by these equations:

$$Ne_{2r}(s, u) = (-1)^r \sqrt{\frac{\pi}{2}} \sum_{k=0}^{\infty} (-1)^k \frac{De_{2k}}{De_0} \left[ Y_k(y) \cdot J_k(x) \right] \tag{14}$$

$$Ne_{2r+1}(s, u) = (-1)^r \sqrt{\frac{\pi}{2}} \sum_{k=0}^{\infty} (-1)^k \frac{De_{2k+1}}{De_1} \left[ Y_{k+1}(y) \cdot J_k(x) + Y_k(y) \cdot J_{k+1}(x) \right] \tag{15}$$

$$No_{2r}(s, u) = (-1)^r \sqrt{\frac{\pi}{2}} \sum_{k=1}^{\infty} (-1)^k \frac{Do_{2k}}{Do_2} \left[ Y_{k+1}(y) \cdot J_{k-1}(x) - Y_{k-1}(y) \cdot J_{k+1}(x) \right] \tag{16}$$

$$No_{2r+1}(s, u) = (-1)^r \sqrt{\frac{\pi}{2}} \sum_{k=0}^{\infty} (-1)^k \frac{Do_{2k+1}}{Do_1} \left[ Y_{k+1}(y) \cdot J_k(x) - Y_k(y) \cdot J_{k+1}(x) \right] \tag{17}$$

where $y = \frac{\sqrt{s}}{2} e^u$, $x = \frac{\sqrt{s}}{2} e^{-u}$, and $J_m(x)$ and $Y_m(y)$ are the Bessel functions of the first and second kinds, respectively. Except when $s = 0$, these series converge rapidly for real values of $u \geq 0$, which is the range permitted in the elliptic coordinate system. However, in spite of satisfactory convergence, Equations 14 through 17 do not lend themselves to easy calculation because of the large variety of Bessel functions involved, not all of which are available in published tables. Part of the difficulty occurs because the arguments of the $J_j(x) \cdot Y_k(y)$ products are cross-linked; if, with a specified s, a value of u is selected which gives a calculated x for which $J_j(x)$ appears in some table, then the corresponding value of y is very often one for which the $Y_k(y)$ needed is not tabulated. As a result of this it was necessary during the preparation of these tables to calculate many specific values of the Bessel functions. Alternatively in some instances, the variable u was adjusted so that, for the specified s, the calculated x and y values were those for which the necessary $J_j(x)$ and $Y_k(y)$ could be found in tables. Unfortunately, the latter tactic produces non-uniform intervals in u.

Different series expansions, listed below, were used for the calculation of some of the $Ne_1$ and $No_1$ data for u greater than approximately 1.14.

$$Ne_1(s,u) = \sqrt{\frac{\pi}{2}} \sum_{k=0}^{\infty} (-1)^k De_{2k+1} \cdot Y_{2k+1}(\sqrt{s} \cosh u) \tag{18}$$

$$No_1(s,u) = \sqrt{\frac{\pi}{2}} \tanh u \sum_{k=0}^{\infty} (-1)^k (2k+1) \cdot Do_{2k+1} \cdot Y_{2k+1}(\sqrt{s} \cosh u) \tag{19}$$

These expansions converge slowly for small values of u; additional restrictions on the representations are given in Reference (4), page xxi. In Table III, the use of Equation (18) or (19) for calculation of $Ne_1$ or $No_1$ is indicated by an asterisk and the lack of numbers in the columns headed by x and y. In addition to the results in Table III, Figures 26 through 35 show curves of the functions of the second kind for orders zero through two and for $0 \le u \le 2$. Except when s = 0, the $Ne_n$ and $No_n$ are finite at u = 0.

The choice of normalization for the radial functions of the first and second kinds is such that when u = 0 the following simple relationships exist for any real $s \ne 0$:

$$Je_n(s,0) = \frac{1}{\left[ Ne_n'(s,u) \right]_{u=0}} \tag{20}$$

$$\left[ Je_n'(s,u) \right]_{u=0} = 0 \tag{21}$$

14

$$Jo_n(s,0) = 0 \tag{22}$$

$$\left[ Jo_n'(s,u) \right]_{u=0} = \frac{-1}{No_n(s,0)} \tag{23}$$

where the primes indicate derivatives with respect to u. These equations provide a check on some of the functions at u = 0. The quantities $Je_n$, $Jo_n$, $Ne_n$ and $No_n$, as well as their derivatives, may also be evaluated at u = 0 by the use of certain joining factors described in Reference (4), page xxiii. These values appear in the tables, usually with a larger number of decimal places than the remainder of the data, since the joining factors are available with high accuracy.

Solutions of Equation (5) analogous to the Hankel functions have also been defined, as follows:

$$He_n^{(1)}(s,u) = Je_n + i\, Ne_n \qquad\qquad Ho_n^{(1)}(s,u) = Jo_n + i\, No_n \tag{24}$$

$$He_n^{(2)}(s,u) = Je_n - i\, Ne_n \qquad\qquad Ho_n^{(2)}(s,u) = Jo_n - i\, No_n \tag{25}$$

where Equations (24) and (25) give the radial Mathieu functions of the third and fourth kinds, respectively. Certain problems dealing with propagation in unbounded space require that the field components of the wave decay exponentially for large values of the radial variable. In such cases the parameter s is often a negative real quantity $(-|s|)$, and under these circumstances the appropriate radial solution is the Hankel-like Mathieu function of the third kind. For example, $He_n^{(1)}(-|s|,u)$ has the following asymptotic form for large u:

$$He_n^{(1)}(-|s|,u) \rightarrow \sqrt{\frac{2}{\sqrt{|s|}}} \cdot e^{-i\left(\frac{n+1}{2}\right)\pi} \cdot e^{-\left(\frac{u}{2} + \frac{\sqrt{|s|}}{2} e^u\right)} \tag{26}$$

Table IV and Figures 36 through 40 contain values and curves for $He_n^{(1)}(-|s|, u)$ and $Ho_n^{(1)}(-|s|, u)$ when $0 \leq n \leq 2$ and $0 \leq u \leq 2$. The data were calculated from the following expressions:

$$He_{2r}^{(1)}(-|s|, u) = i(-1)^{r+1}\sqrt{\frac{2}{\pi}}\sum_{m=o}^{\infty}\frac{De_{2m}}{De_o}I_m(x) \cdot K_m(y) \tag{27}$$

$$He_{2r+1}^{(1)}(-|s|, u) = (-1)^{r+1}\sqrt{\frac{2}{\pi}}\sum_{m=o}^{\infty}\frac{Do_{2m+1}}{Do_1}\left[I_m(x) \cdot K_{m+1}(y) - I_{m+1}(x) \cdot K_m(y)\right] \tag{28}$$

$$Ho_{2r}^{(1)}(-|s|, u) = i(-1)^{r}\sqrt{\frac{2}{\pi}}\sum_{m=o}^{\infty}\frac{Do_{2m}}{Do_2}\left[I_{m+1}(x) \cdot K_{m-1}(y) - I_{m-1}(x) \cdot K_{m+1}(y)\right] \tag{29}$$

$$Ho_{2r+1}^{(1)}(-|s|, u) = (-1)^{r+1}\sqrt{\frac{2}{\pi}}\sum_{m=o}^{\infty}\frac{De_{2m+1}}{De_1}\left[I_{m+1}(x) \cdot K_m(y) + I_m(x) \cdot K_{m+1}(y)\right] \tag{30}$$

where $y = \sqrt{\frac{|s|}{2}}e^u$, $x = \sqrt{\frac{|s|}{2}}e^{-u}$, and $I_k(x)$ and $K_k(y)$ are the modified Bessel functions. Except when $s = 0$, the $He_n^{(1)}$ and $Ho_n^{(1)}$ functions are finite at $u = 0$. As a consequence of the fact that the odd coefficients are normalized to make $\sum_{m=o}^{\infty} 2m\, Do_{2m} = 1$, Equation (29) reduces to the following simple form when $u = 0$:

$$Ho_{2r}^{(1)}(-|s|, 0) = i(-1)^{r+1}\frac{4}{|s| \cdot Do_2}\sqrt{\frac{2}{\pi}} \tag{31}$$

Similarly, the even coefficients are normalized to make $\sum_{m=o}^{\infty} De_{2m+1} = 1$, and at $u = 0$, Equation (30) reduces to the form:

$$Ho_{2r+1}^{(1)}(-|s|, 0) = (-1)^{r+1}\frac{2}{De_1}\sqrt{\frac{2}{\pi \cdot |s|}} \tag{32}$$

It should be noted that, in general, as s goes to zero $(\text{i.e.}, \stackrel{e,g}{,}$ as the ellipse becomes a circle) the various radial Mathieu functions reduce to the Bessel functions of the same kind multiplied by a constant. For example, as $s \to 0$, $Je_n(s, u)$ or $Jo_n(s, u) \to \sqrt{\frac{\pi}{2}}\ J_n(k\rho)$, where $\rho$ is the radial variable of the circular cylinder coordinate system and k is the constant used earlier in the wave equation.

The tables included in this report contain a total of approximately 1750 calculated values of the various Mathieu functions. The radial functions were not calculated at uniform intervals in the variable u, mainly because the presently available tables of the Bessel functions are inadequate (see earlier comments). Also intervals in u were often chosen smaller near maxima, minima, or zeros of the functions, since such points are frequently of most interest in physical problems. The positions of maxima and minima of all radial functions were also checked for the cases s = 3, 6, or 9 by finding the points at which the derivatives of the functions vanished. The functions of the third kind, which are monotonic and have non-vanishing derivatives for $0 \le u < \infty$, were checked roughly (for s = 3, 6 or 9) by comparing the slope of the function curves against values of the derivatives at several points in the region $0 \le u \le 2$.

The plotted curves should have considerable value because they not only show the stationary and zero points of the functions, but also permit interpolation of the tabulated values and give an indication of the effect of variation of the parameter s. Points read from the curves, however, have accuracy to a smaller number of decimal places than do the values listed in the tables. The accuracy of the latter data is such that the estimated maximum error of any tabular value is plus or minus one unit in the final decimal place.

17

# REFERENCES

1.  McLachlan, N. W., Theory and Application of Mathieu Functions, University Press, Oxford, 1951.

2.  Brillouin, L., Wave Propagation in Periodic Structures, Second Edition, Dover Publications, Inc., New York, 1953.

3.  Morse, P. M., and Feshbach, H., Methods of Theoretical Physics, McGraw-Hill Book Company, Inc., New York, 1953.

4.  National Bureau of Standards, Tables Relating to Mathieu Functions, Columbia University Press, New York, 1951.

5.  Stratton, J. A., Morse, P. M., Chu, L. J., and Hutner, R. A., Elliptic Cylinder and Spheroidal Wave Functions, John Wiley and Sons, Inc., New York, 1941.

6.  Stratton, J. A., Electromagnetic Theory, McGraw-Hill Book Company, Inc., New York, 1941.

# BIBLIOGRAPHY

Blanch, G. and Clemm, D.S., TABLES RELATING TO THE RADIAL MATHIEU FUNCTIONS, Vol 1, Functions of the First Kind (1963), and Vol. 2, Functions of the Second Kind (1964), Aeronautical Research Laboratories, United States air Force, U.S. Government Printing Office, Washignton 25, D.C.

Chu, L.J., ELECTROMAGNETIC WAVES IN ELLIPTICAL HOLLOW PIPES OF METAL, Journal of Applied Physics, Vol. 9, p. 583, september, 1938.

Dyott, R.B., ELLIPTICAL FIBER WAVEQUIDES, Artech House, Norwood, MA, 1995.

Kornhauser, E.T., ON THE DISCRETE SPECTRUM FOR DIELECTRIC RODS OF ARBITRARY CROSS SECTION, Brown Univ. Scientific Rept AF 451/3, March, 1959. (ASTIA document No. AD 211 930).

National Bureau of Standards, TABLES RELATING TO THE MATHIEU FUNCTIONS, Superintendent of Documents, U.S. Government Printing Office, 1967.

# TABLE I

## Periodic Mathieu Functions

$$Se_0(s,v) = \sum_{k=0}^{\infty} De_{2k} \cdot \cos 2kv$$

| | s = 1 | 2 | 3 | 4 | 5 | 6 |
|---|---|---|---|---|---|---|
| v = 0° | 1.0000 | 1.0000 | 1.0000 | 1.0000 | 1.0000 | 1.0000 |
| 5° | 1.0020 | 1.0043 | 1.0067 | 1.0094 | 1.0121 | 1.0150 |
| 15° | 1.0179 | 1.0379 | 1.0599 | 1.0837 | 1.1090 | 1.1355 |
| 25° | 1.0480 | 1.1025 | 1.1633 | 1.2299 | 1.3016 | 1.3780 |
| 30° | | | 1.2319 | | | 1.5444 |
| 35° | 1.0892 | 1.1926 | 1.3099 | 1.4406 | 1.5840 | 1.7395 |
| 45° | 1.1370 | 1.2992 | 1.4869 | 1.7004 | 1.9391 | 2.2029 |
| 55° | 1.1859 | 1.4104 | 1.6756 | 1.9828 | 2.3329 | 2.7268 |
| 60° | | | 1.7667 | | | 2.9906 |
| 65° | 1.2298 | 1.5121 | 1.8514 | 2.2509 | 2.7133 | 3.2415 |
| 75° | 1.2629 | 1.5901 | 1.9881 | 2.4623 | 3.0175 | 3.6585 |
| 85° | 1.2807 | 1.6324 | 2.0631 | 2.5794 | 3.1873 | 3.8931 |
| 90° | | | 2.0728 | | | 3.9238 |

| | s = 7 | 8 | 9 | 15 | 25 | 40 |
|---|---|---|---|---|---|---|
| v = 0° | 1.0000 | 1.0000 | 1.0000 | 1.0000 | 1.0000 | 1.0000 |
| 5° | 1.0180 | 1.0210 | 1.0241 | | | |
| 15° | 1.1631 | 1.1916 | 1.2209 | 1.4102 | 1.7657 | 2.3816 |
| 25° | 1.4585 | 1.5429 | 1.6310 | | | |
| 30° | | | 1.9226 | 2.8510 | 4.9233 | 9.5170 |
| 35° | 1.9065 | 2.0846 | 2.2737 | | | |
| 45° | 2.4917 | 2.8055 | 3.1448 | 5.7539 | 12.7990 | 32.8353 |
| 55° | 3.1656 | 3.6508 | 4.1841 | | | |
| 60° | | | 4.7268 | 10.0706 | 26.9696 | 85.5843 |
| 65° | 3.8385 | 4.5080 | 5.2541 | | | |
| 75° | 4.3906 | 5.2198 | 6.1527 | 14.3840 | 43.2116 | 156.6422 |
| 85° | 4.7036 | 5.6264 | 6.6698 | | | |
| 90° | | | 6.7379 | 16.2532 | 50.7692 | 192.5729 |

$$Se_1(s,v) = \sum_{k=o}^{\infty} De_{2k+1} \cdot \cos(2k+1)v$$

| | s = 3 | 6 | 9 | 15 | 25 | 40 |
|---|---|---|---|---|---|---|
| v = 0° | 1.0000 | 1.0000 | 1.0000 | 1.0000 | 1.0000 | 1.0000 |
| 5° | .9993 | 1.0032 | 1.0078 | | | |
| 15° | .9929 | 1.0264 | 1.0675 | 1.1738 | 1.4079 | 1.8495 |
| 25° | .9748 | 1.0622 | 1.1728 | | | |
| 30° | .9585 | 1.0786 | 1.2335 | 1.6650 | 2.7555 | 5.3056 |
| 35° | .9355 | 1.0896 | 1.2922 | | | |
| 45° | .8632 | 1.0788 | 1.3745 | 2.2796 | 4.9716 | 12.9283 |
| 55° | .7474 | .9975 | 1.3545 | | | |
| 60° | .6711 | .9220 | 1.2866 | 2.5036 | 6.6778 | 21.6477 |
| 65° | .5825 | .8209 | 1.1730 | | | 22.8176 |
| 75° | .3716 | .5444 | .8058 | 1.7315 | 5.2220 | 19.4020 |
| 85° | .1279 | .1911 | .2880 | | | 7.7229 |
| 90° | .0000 | .0000 | .0000 | .0000 | .0000 | .0000 |

$$Se_2(s,v) = \sum_{k=o}^{\infty} De_{2k} \cos 2kv$$

| | s = 3 | 6 | 9 | 15 | 25 | 40 |
|---|---|---|---|---|---|---|
| v = 0° | 1.0000 | 1.0000 | 1.0000 | 1.0000 | 1.0000 | 1.0000 |
| 5° | .9896 | .9933 | .9966 | | | |
| 15° | .9072 | .9385 | .9661 | 1.0235 | 1.1499 | 1.4326 |
| 25° | .7444 | .8220 | .8909 | | | |
| 30° | .6348 | .7379 | .8294 | 1.0256 | 1.5005 | 2.7636 |
| 35° | .5086 | .6356 | .7481 | | | |
| 45° | .2165 | .3787 | .5194 | .8247 | 1.6450 | 4.3248 |
| 55° | -.1023 | .0693 | .2086 | | | |
| 60° | -.2589 | -.0929 | .0338 | .2782 | .9499 | 3.6125 |
| 65° | -.4058 | -.2506 | -.1428 | | | |
| 75° | -.6456 | -.5193 | -.4570 | -.4287 | -.5279 | -.7564 |
| 85° | -.7784 | -.6738 | -.6444 | | | |
| 90° | -.7956 | -.6942 | -.6695 | -.7687 | -1.3477 | -3.7690 |

$$So_1(s,v) = \sum_{k=0}^{\infty} Do_{2k+1} \cdot \sin(2k+1)\,v$$

|  | s = 3 | 6 | 9 | 15 | 25 | 40 |
|---|---|---|---|---|---|---|
| v = 0° | .0000 | .0000 | .0000 | .0000 | .0000 | .0000 |
| 5° | .0874 | .0877 | .0880 | | | |
| 15° | .2656 | .2727 | .2803 | .2962 | .3250 | .3727 |
| 25° | .4524 | .4849 | .5201 | | | |
| 30° | .5498 | .6051 | .6660 | .8039 | 1.0840 | 1.6379 |
| 35° | .6495 | .7355 | .8319 | | | |
| 45° | .8524 | 1.0241 | 1.2242 | 1.7186 | 2.8778 | 5.6849 |
| 55° | 1.0500 | 1.3340 | 1.6779 | | | |
| 60° | 1.1415 | 1.4869 | 1.9122 | 3.0505 | 6.0845 | 14.8261 |
| 65° | 1.2249 | 1.6310 | 2.1388 | | | |
| 75° | 1.3571 | 1.8686 | 2.5235 | 4.3751 | 9.7562 | 27.1383 |
| 85° | 1.4285 | 2.0015 | 2.7442 | | | |
| 90° | 1.4378 | 2.0188 | 2.7733 | 4.9485 | 11.4643 | 33.3547 |

$$So_2(s,v) = \sum_{k=1}^{\infty} Do_{2k} \cdot \sin 2kv$$

|  | s = 3 | 6 | 9 | 15 | 25 | 40 |
|---|---|---|---|---|---|---|
| v = 0° | .0000 | .0000 | .0000 | .0000 | .0000 | .0000 |
| 5° | .0870 | .0872 | .0874 | | | |
| 15° | .2544 | .2590 | .2640 | .2749 | .2954 | .3312 |
| 25° | .4010 | .4206 | .4421 | | | |
| 30° | .4616 | .4934 | .5285 | .6094 | .7782 | 1.1194 |
| 35° | .5110 | .5573 | .6094 | | | |
| 45° | .5676 | .6463 | .7375 | .9628 | 1.4934 | 2.7797 |
| 55° | .5565 | .6606 | .7850 | | | |
| 60° | .5230 | .6326 | .7654 | 1.1161 | 2.0404 | 4.6704 |
| 65° | .4708 | .5791 | .7120 | 1.0694 | | 4.9249 |
| 75° | .3158 | .3986 | .5021 | .7887 | 1.6047 | 4.1897 |
| 85° | .1113 | .1424 | .1816 | | | |
| 90° | .0000 | .0000 | .0000 | .0000 | .0000 | .0000 |

$$Se_o(-|s|,v) = \frac{1}{Se_o(|s|,\frac{\pi}{2})} \sum_{k=o}^{\infty} (-1)^k De_{2k} \cdot \cos 2kv$$

|  | s = 3 | 6 | 9 | 15 |
|---|---|---|---|---|
| v = 0° | 1.0000 | 1.0000 | 1.0000 | 1.0000 |
| 15° | .9592 | .9324 | .9131 | .8850 |
| 30° | .8523 | .7622 | .7015 | .6196 |
| 45° | .7174 | .5614 | .4667 | .3540 |
| 60° | .5943 | .3936 | .2853 | .1754 |
| 75° | .5114 | .2894 | .1812 | .0868 |
| 90° | .4824 | .2549 | .1484 | .0615 |

$$Se_1(-|s|,v) = \frac{1}{So_1(|s|,\frac{\pi}{2})} \sum_{k=o}^{\infty} (-1)^k Do_{2k+1} \cdot \cos (2k+1) v$$

|  | s = 3 | 6 | 9 | 15 | 25 |
|---|---|---|---|---|---|
| v = 0° | 1.0000 | 1.0000 | 1.0000 | 1.0000 | 1.0000 |
| 15° | .9439 | .9256 | .9099 | .8841 | .8510 |
| 30° | .7940 | .7365 | .6895 | .6164 | .5308 |
| 45° | .5929 | .5073 | .4414 | .3473 | .2510 |
| 60° | .3824 | .2997 | .2401 | .1624 | .0946 |
| 75° | .1847 | .1351 | .1011 | .0599 | .0284 |
| 90° | .0000 | .0000 | .0000 | .0000 | .0000 |

$$So_1(-|s|,v) = \frac{1}{Se_1'(|s|,\frac{\pi}{2})} \sum_{k=o}^{\infty} (-1)^{k+1} De_{2k+1} \cdot \sin (2k+1)v$$

|  | s = 3 | 6 | 9 | 15 | 25 |
|---|---|---|---|---|---|
| v = 0° | .0000 | .0000 | .0000 | .0000 | .0000 |
| 15° | .2526 | .2470 | .2420 | .2340 | .2246 |
| 30° | .4561 | .4183 | .3865 | .3384 | .2872 |
| 45° | .5867 | .4894 | .4128 | .3081 | .2138 |
| 60° | .6515 | .4893 | .3705 | .2251 | .1185 |
| 75° | .6748 | .4656 | .3206 | .1587 | .0605 |
| 90° | .6797 | .4536 | .3004 | .1352 | .0430 |

## TABLE II

### Radial Mathieu Functions of the First Kind

$$Je_o(s,u) = \sqrt{\frac{\pi}{2}} \sum_{k=0}^{\infty} (-1)^k \, De_{2k} \cdot J_{2k}(x) \; ; \; x = \sqrt{s} \, \cosh u$$

| s = 1 | | | s = 3 | | |
|---|---|---|---|---|---|
| u | $Je_o(s,u)$ | x | u | $Je_o(s,u)$ | x |
| .0000 | 1.11296 | | .0000 | .91402 | |
| .4436 | 1.0517 | 1.10 | .1441 | .8973 | 1.75 |
| .6224 | .9865 | 1.20 | .2792 | .8505 | 1.80 |
| .7564 | .9179 | 1.30 | .3670 | .8035 | 1.85 |
| .8671 | .8462 | 1.40 | .5493 | .6622 | 2.00 |
| 1.0470 | .6958 | 1.60 | .9100 | .2102 | 2.50 |
| 1.1929 | .5398 | 1.80 | 1.0371 | .0105 | 2.75 |
| 1.3169 | .3823 | 2.00 | 1.1462 | -.1621 | 3.00 |
| 1.4254 | .2278 | 2.20 | 1.1861 | -.2222 | 3.10 |
| 1.5221 | .0802 | 2.40 | 1.3289 | -.4058 | 3.50 |
| 1.5668 | .0102 | 2.50 | 1.4796 | -.4991 | 4.00 |
| 1.6892 | -.1792 | 2.80 | 1.6086 | -.4501 | 4.50 |
| 1.7628 | -.2848 | 3.00 | 1.7218 | -.2930 | 5.00 |
| 1.8946 | -.4364 | 3.40 | 1.8228 | -.0793 | 5.50 |
| 2.0099 | -.5015 | 3.80 | 1.9974 | +.2950 | 6.50 |

| s = 6 | | | s = 6 | | |
|---|---|---|---|---|---|
| u | $Je_o(3,u)$ | x | u | $Je_o(3,u)$ | x |
| .0000 | .74497 | | 1.3404 | -.3698 | 5.00 |
| .0205 | .7444 | 2.45 | 1.4483 | -.1686 | 5.50 |
| .4904 | .4018 | 2.75 | 1.5204 | -.0007 | 5.87 |
| .6585 | .1488 | 3.00 | 1.6315 | .2456 | 6.50 |
| .7393 | .0144 | 3.15 | 1.7111 | .3548 | 7.00 |
| .8959 | -.2406 | 3.50 | 1.8524 | .2806 | 8.00 |
| 1.0730 | -.4502 | 4.00 | 2.0 | -.113 | 9.22 |
| 1.1905 | -.4888 | 4.40 | | | |

$$Je_o(s,u) = \sqrt{\frac{\pi}{2}} \sum_{k=o}^{\infty} (-1)^k De_{2k} \cdot J_{2k}(x), \quad \text{(cont'd)}$$

| s = 9 | | | s = 15 | | |
|---|---|---|---|---|---|
| u | $Je_o(s,u)$ | x | u | $Je_o(s,u)$ | x |
| .0000 | .65217 | | .0000 | .55387 | |
| .3045 | .4665 | 3.14 | .0602 | .542 | 3.88 |
| .5697 | .0597 | 3.50 | .2554 | .358 | 4.00 |
| .6020 | .0026 | 3.56 | .4490 | .015 | 4.27 |
| .7954 | −.3168 | 4.00 | .5616 | −.199 | 4.50 |
| .9888 | −.4766 | 4.59 | .7456 | −.435 | 5.00 |
| 1.0986 | −.4331 | 5.00 | .8087 | −.453 | 5.21 |
| 1.2149 | −.2636 | 5.50 | 1.0052 | −.238 | 6.00 |
| 1.3170 | −.0357 | 6.00 | 1.1073 | .002 | 6.50 |
| 1.4082 | .1787 | 6.50 | 1.1979 | .215 | 7.00 |
| 1.4910 | .3239 | 7.00 | 1.3540 | .354 | 8.00 |
| 1.5697 | .3688 | 7.52 | 1.5131 | .028 | 9.22 |
| 1.6368 | .3126 | 8.00 | 1.6019 | −.218 | 10.00 |
| 1.7627 | .0053 | 9.00 | 1.6505 | −.295 | 10.46 |
| 1.7883 | −.0704 | 9.22 | 1.7045 | −.295 | 11.00 |
| 1.8738 | −.2718 | 10.00 | 1.7969 | −.080 | 12.00 |
| 1.9733 | −.2681 | 11.00 | 1.8811 | .235 | 13.00 |
| | | | 1.9585 | .262 | 14.00 |

| s = 25 | | | s = 25 | | |
|---|---|---|---|---|---|
| u | $Je_o(s,u)$ | x | u | $Je_o(s,u)$ | x |
| .000 | .47544 | | 1.1929 | .262 | 9.00 |
| .3390 | .014 | 5.29 | 1.3169 | −.082 | 10.00 |
| .4436 | −.206 | 5.50 | 1.4254 | −.298 | 11.00 |
| .6224 | −.419 | 6.00 | 1.5221 | −.182 | 12.00 |
| .7141 | −.381 | 6.33 | 1.6094 | .108 | 13.00 |
| .7564 | −.326 | 6.50 | 1.6892 | .267 | 14.00 |
| .8671 | −.089 | 7.00 | 1.7628 | .152 | 15.00 |
| 1.0470 | .314 | 8.00 | 1.8309 | −.102 | 16.00 |
| 1.1203 | .348 | 8.48 | 1.9542 | −.142 | 18.00 |

$$Je_0(s,u) = \sqrt{\frac{\pi}{2}} \sum_{k=o}^{\infty} (-1)^k De_{2k} \cdot J_{2k}(x) , \quad (\text{cont'd})$$

| s = 40 | | | s = 40 | | |
|---|---|---|---|---|---|
| u | $Je_0(s,u)$ | x | u | $Je_0(s,u)$ | x |
| .0000 | .41625 |  | 1.2556 | -.283 | 12.00 |
| .0415 | .404 | 6.33 | 1.3484 | -.051 | 13.00 |
| .2350 | .080 | 6.50 | 1.4323 | .215 | 14.00 |
| .4581 | -.363 | 7.00 | 1.5090 | .234 | 15.00 |
| .5202 | -.385 | 7.20 | 1.5797 | .010 | 16.00 |
| .6006 | -.314 | 7.50 | 1.6454 | -.209 | 17.00 |
| .7127 | -.060 | 8.00 | 1.7067 | -.205 | 18.00 |
| .8073 | .191 | 8.50 | 1.7642 | .002 | 19.00 |
| .9207 | .336 | 9.20 | 1.8185 | .195 | 20.00 |
| 1.0317 | .174 | 10.00 | 1.8697 | .189 | 21.00 |
| 1.1513 | -.184 | 11.00 | 1.9184 | -.000 | 22.00 |

$$Je_1(s,u) = \sqrt{\frac{\pi}{2}} \sum_{k=o}^{\infty} (-1)^k De_{2k+1} \cdot J_{2k+1}(x) ; \quad x = \sqrt{s} \cosh u$$

| s = 1 | | | s = 3 | | |
|---|---|---|---|---|---|
| u | $Je_1(s,u)$ | x | u | $Je_1(s,u)$ | x |
| .0000 | .57051 |  | .0000 | .81918 |  |
| .4436 | .6107 | 1.10 | .1441 | .8206 | 1.75 |
| .6224 | .6465 | 1.20 | .5493 | .8211 | 2.00 |
| .8671 | .7037 | 1.40 | .9100 | .7228 | 2.50 |
| 1.1232 | .7516 | 1.70 | 1.1462 | .5163 | 3.00 |
| 1.3169 | .7520 | 2.00 | 1.3289 | .2468 | 3.50 |
| 1.5668 | .6526 | 2.50 | 1.4796 | -.0298 | 4.00 |
| 1.6892 | .5418 | 2.80 | 1.6086 | -.2600 | 4.50 |
| 1.7628 | .4519 | 3.00 | 1.7218 | -.4025 | 5.00 |
| 1.8309 | .3527 | 3.20 | 1.8228 | -.4371 | 5.50 |
| 1.8946 | .2476 | 3.40 | 1.9974 | -.2174 | 6.50 |
| 1.9542 | .1403 | 3.60 |  |  |  |
| 1.9827 | .0868 | 3.70 |  |  |  |
| 1.9994 | .0551 | 3.76 |  |  |  |
| 2.0099 | .0341 | 3.80 |  |  |  |
| 2.1 | -.1360 | 4.144 |  |  |  |

$$Je_1(s,u) = \sqrt{\frac{\pi}{2}} \ \sum_{k=0}^{\infty} (-1)^k \ De_{2k+1} \cdot J_{2k+1}(x), \quad \text{(cont'd)}$$

| | s = 6 | | | s = 9 | |
|---|---|---|---|---|---|
| u | $Je_1(s,u)$ | x | u | $Je_1(s,u)$ | x |
| .0000 | .87975 | | .0000 | .83255 | |
| .0205 | .8796 | 2.45 | .3045 | .7496 | 3.14 |
| .4904 | .7681 | 2.75 | .5697 | .5134 | 3.50 |
| .6585 | .6469 | 3.00 | .7954 | .1695 | 4.00 |
| .8959 | .3558 | 3.50 | .8904 | -.0035 | 4.27 |
| 1.0730 | .0502 | 4.00 | 1.0987 | -.3481 | 5.00 |
| 1.3404 | -.3840 | 5.00 | 1.2149 | -.4412 | 5.50 |
| 1.4483 | -.4428 | 5.50 | 1.3170 | -.4115 | 6.00 |
| 1.6315 | -.2473 | 6.50 | 1.4083 | -.2823 | 6.50 |
| 1.7111 | -.0582 | 7.00 | 1.4910 | -.0951 | 7.00 |
| 1.8524 | .2734 | 8.00 | 1.6368 | .2555 | 8.00 |
| 1.9305 | .3452 | 8.62 | 1.7206 | .3452 | 8.65 |
| 2.0 | .290 | 9.22 | 1.7627 | .3318 | 9.00 |
| | | | 1.7883 | .3026 | 9.22 |
| | | | 1.8738 | .1040 | 10.00 |
| | | | 1.9733 | -.1928 | 11.00 |

| | s = 15 | | | s = 15 | |
|---|---|---|---|---|---|
| u | $Je_1(s,u)$ | x | u | $Je_1(s,u)$ | x |
| .0000 | .69851 | | 1.2606 | -.0342 | 7.38 |
| .0602 | .6920 | 3.88 | 1.3540 | .1982 | 8.0 |
| .2554 | .5820 | 4.00 | 1.4602 | .3434 | 8.79 |
| .5617 | .1504 | 4.50 | 1.5131 | .3268 | 9.22 |
| .6386 | .0078 | 4.69 | 1.6019 | .1531 | 10.00 |
| .7456 | -.1882 | 5.00 | 1.6505 | .0057 | 10.46 |
| .9602 | -.4381 | 5.80 | 1.7045 | -.1573 | 11.00 |
| 1.0052 | -.4411 | 6.00 | 1.7969 | -.2937 | 12.00 |
| 1.1073 | -.3604 | 6.50 | 1.8811 | -.1504 | 13.00 |
| 1.1979 | -.1898 | 7.00 | 1.9585 | .1183 | 14.00 |

$$Je_1(s,u) = \sqrt{\frac{\pi}{2}} \sum_{k=0}^{\infty} (-1)^k De_{2k+1} \cdot J_{2k+1}(x) , \quad \text{(cont'd)}$$

| | s = 25 | | | s = 40 | |
|---|---|---|---|---|---|
| u | $Je_1(s,u)$ | x | u | $Je_1(s,u)$ | x |
| .0000 | .55925 | | .0000 | .46523 | |
| .4436 | .014 | 5.50 | .0415 | .456 | 6.33 |
| .6224 | -.315 | 6.00 | .2350 | .202 | 6.50 |
| .7564 | -.427 | 6.50 | .3290 | -.003 | 6.67 |
| .8671 | -.362 | 7.00 | .4581 | -.272 | 7.00 |
| 1.0470 | .034 | 8.00 | .6006 | -.401 | 7.50 |
| 1.1929 | .326 | 9.00 | .7127 | -.295 | 8.00 |
| 1.2218 | .341 | 9.22 | .8073 | -.076 | 8.50 |
| 1.3169 | .245 | 10.00 | .9237 | .225 | 9.22 |
| 1.4254 | -.068 | 11.00 | 1.0317 | .331 | 10.00 |
| 1.5221 | -.283 | 12.00 | 1.1513 | .110 | 11.00 |
| 1.6094 | -.208 | 13.00 | 1.2556 | -.205 | 12.00 |
| 1.6892 | .057 | 14.00 | 1.3484 | -.274 | 13.00 |
| 1.7628 | .248 | 15.00 | 1.4323 | -.058 | 14.00 |
| 1.8309 | .194 | 16.00 | 1.5090 | .198 | 15.00 |
| 1.8946 | -.037 | 17.00 | 1.5797 | .240 | 16.00 |
| 1.9542 | -.220 | 18.00 | 1.6454 | .046 | 17.00 |
| | | | 1.7067 | -.180 | 18.00 |
| | | | 1.8185 | -.049 | 20.00 |
| | | | 1.9184 | .208 | 22.00 |

$$Je_2(s,u) = -\sqrt{\frac{\pi}{2}} \sum_{k=0}^{\infty} (-1)^k De_{2k} \cdot J_{2k}(x); \quad x = \sqrt{s} \cosh u$$

| | s = 1 | | | s = 1 | |
|---|---|---|---|---|---|
| u | $Je_2(s,u)$ | x | u | $Je_2(s,u)$ | x |
| .0000 | .08116 | | 1.4254 | .4684 | 2.20 |
| .4436 | .1106 | 1.10 | 1.5221 | .5201 | 2.40 |
| .6224 | .1418 | 1.20 | 1.5668 | .5423 | 2.50 |
| .7564 | .1741 | 1.30 | 1.6892 | .5914 | 2.80 |
| .8671 | .2074 | 1.40 | 1.7628 | .6078 | 3.00 |
| 1.0470 | .2756 | 1.60 | 1.8946 | .5973 | 3.40 |
| 1.1929 | .3436 | 1.80 | 2.0099 | .5290 | 3.80 |
| 1.3169 | .4088 | 2.00 | | | |

$$Je_2(s,u) = -\sqrt{\frac{\pi}{2}} \sum_{k=o}^{\infty} (-1)^k \, De_{2k} \cdot J_{2k}(x); \qquad \text{(cont'd)}$$

| | s = 3 | | | s = 6 | |
| --- | --- | --- | --- | --- | --- |
| u | $Je_2(s,u)$ | x | u | $Je_2(s,u)$ | x |
| .0000 | .25105 | | .0000 | .48020 | |
| .1441 | .2581 | 1.75 | .0205 | .4804 | 2.45 |
| .2792 | .2777 | 1.80 | .4904 | .5690 | 2.75 |
| .3670 | .2972 | 1.85 | .6585 | .6182 | 3.00 |
| .5493 | .3541 | 2.00 | .8959 | .6396 | 3.50 |
| .9100 | .5161 | 2.50 | 1.0730 | .5566 | 4.00 |
| 1.0371 | .5725 | 2.75 | 1.3404 | .1646 | 5.00 |
| 1.1462 | .6076 | 3.00 | 1.4483 | -.0641 | 5.50 |
| 1.1861 | .6151 | 3.10 | 1.6315 | -.3700 | 6.50 |
| 1.3289 | .6059 | 3.50 | 1.7111 | -.3933 | 7.00 |
| 1.4796 | .5086 | 4.00 | 1.8524 | -.1890 | 8.00 |
| 1.6086 | .3339 | 4.50 | 2.0 | .207 | 9.22 |
| 1.7218 | .1160 | 5.00 | | | |
| 1.8228 | -.1017 | 5.50 | | | |
| 1.9974 | -.3772 | 6.50 | | | |

| | s = 9 | | | s = 9 | |
| --- | --- | --- | --- | --- | --- |
| u | $Je_2(s,u)$ | x | u | $Je_2(s,u)$ | x |
| .0000 | .64181 | | 1.4910 | -.3987 | 7.00 |
| .5697 | .6814 | 3.50 | 1.6368 | -.2078 | 8.00 |
| .7954 | .6088 | 4.00 | 1.7094 | -.0150 | 8.56 |
| 1.0987 | .2132 | 5.00 | 1.7627 | .1326 | 9.00 |
| 1.2149 | -.0268 | 5.50 | 1.7883 | .1957 | 9.22 |
| 1.3170 | -.2314 | 6.00 | 1.8738 | .3191 | 10.00 |
| 1.4083 | -.3624 | 6.50 | 1.9733 | .2103 | 11.00 |

$$Je_2(s,u) = -\sqrt{\frac{\pi}{2}} \sum_{k=o}^{\infty} (-1)^k De_{2k} \cdot J_{2k}(x), \qquad \text{(cont'd)}$$

| s = 15 | | | s = 25 | | |
|---|---|---|---|---|---|
| u | $Je_2(s,u)$ | x | u | $Je_2(s,u)$ | x |
| .0000 | .77140 | | .0000 | .70576 | |
| .0602 | .7702 | 3.88 | .3022 | .5528 | 5.23 |
| .2554 | .7459 | 4.00 | .4436 | .3665 | 5.50 |
| .5617 | .5787 | 4.50 | .6224 | .0364 | 6.00 |
| .7456 | .3357 | 5.00 | .7141 | -.1469 | 6.33 |
| .9099 | .0212 | 5.59 | .7564 | -.2251 | 6.50 |
| 1.0052 | -.1714 | 6.00 | .8671 | -.3778 | 7.00 |
| 1.1073 | -.3377 | 6.50 | .9201 | -.4088 | 7.27 |
| 1.1979 | -.4047 | 7.00 | 1.0470 | -.3254 | 8.00 |
| 1.3540 | -.2472 | 8.00 | 1.1929 | .0208 | 9.00 |
| 1.5131 | .1694 | 9.22 | 1.2218 | .0998 | 9.22 |
| 1.6019 | .3163 | 10.00 | 1.3169 | .2974 | 10.00 |
| 1.6505 | .3139 | 10.46 | 1.3696 | .3251 | 10.47 |
| 1.7045 | .2303 | 11.00 | 1.4254 | .2684 | 11.00 |
| 1.7969 | -.0502 | 12.00 | 1.5221 | -.0016 | 12.00 |
| 1.8811 | -.2614 | 13.00 | 1.6094 | -.2445 | 13.00 |
| 1.9585 | -.2235 | 14.00 | 1.6892 | -.2450 | 14.00 |
| | | | 1.7628 | -.0236 | 15.00 |
| | | | 1.8309 | .2033 | 16.00 |
| | | | 1.8946 | .2311 | 17.00 |
| | | | 1.9542 | .0479 | 18.00 |

| s = 40 | | | s = 40 | | |
|---|---|---|---|---|---|
| u | $Je_2(s,u)$ | x | u | $Je_2(s,u)$ | x |
| .0000 | .55381 | | 1.3484 | -.1858 | 13.00 |
| .0415 | .5479 | 6.33 | 1.4323 | -.2736 | 14.00 |
| .2350 | .3716 | 6.50 | 1.5090 | -.0956 | 15.00 |
| .4581 | -.0575 | 7.00 | 1.5797 | .1586 | 16.00 |
| .6006 | -.3196 | 7.50 | 1.6454 | .2468 | 17.00 |
| .7127 | -.4047 | 8.00 | 1.7067 | .0998 | 18.00 |
| .8073 | -.3387 | 8.50 | 1.7642 | -.1309 | 19.00 |
| .9237 | -.0856 | 9.22 | 1.8185 | -.2274 | 20.00 |
| 1.0317 | .2022 | 10.00 | 1.8697 | -.1088 | 21.00 |
| 1.1513 | .3180 | 11.00 | 1.9184 | .1040 | 22.00 |
| 1.2556 | .1065 | 12.00 | | | |

$$Jo_1(s,u) = \sqrt{\tfrac{\pi}{2}}\ \tanh u \sum_{k=0}^{\infty} (-1)^k (2k+1)\, Do_{2k+1}\cdot J_{2k+1}(x); \quad x = \sqrt{s}\,\cosh u$$

| s = 1 | | | | s = 3 | | |
|---|---|---|---|---|---|---|
| u | $Jo_1(s,u)$ | x | | u | $Jo_1(s,u)$ | x |
| .0000 | .0000 | | | .0000 | .0000 | |
| .4436 | .271 | 1.10 | | .0959 | .0955 | 1.74 |
| .6224 | .381 | 1.20 | | .1441 | .1431 | 1.75 |
| .8671 | .526 | 1.40 | | .2792 | .2736 | 1.80 |
| 1.1232 | .652 | 1.70 | | .3670 | .3545 | 1.85 |
| 1.3169 | .701 | 2.00 | | .5493 | .5050 | 2.00 |
| 1.5668 | .652 | 2.50 | | .8523 | .6548 | 2.40 |
| 1.6892 | .559 | 2.80 | | ..9100 | .6602 | 2.50 |
| 1.7628 | .476 | 3.00 | | .9635 | .6564 | 2.60 |
| 1.8309 | .382 | 3.20 | | 1.0371 | .6352 | 2.75 |
| 1.8946 | .281 | 3.40 | | 1.1462 | .5663 | 3.00 |
| 1.9542 | .174 | 3.60 | | 1.1861 | .5290 | 3.10 |
| 1.9827 | .121 | 3.70 | | 1.3156 | .3624 | 3.46 |
| 2.0099 | .068 | 3.80 | | 1.3289 | .3417 | 3.50 |
| | | | | 1.4796 | .0683 | 4.00 |
| | | | | 1.5204 | -.0123 | 4.15 |
| | | | | 1.5668 | -.1039 | 4.33 |
| | | | | 1.6086 | -.1832 | 4.50 |
| | | | | 1.7218 | -.3587 | 5.00 |
| | | | | 1.7636 | -.3997 | 5.20 |
| | | | | 1.8228 | -.4272 | 5.50 |
| | | | | 1.9245 | -.3729 | 6.06 |
| | | | | 1.9974 | -.2523 | 6.50 |
| | | | | 2.063 | -.0960 | 6.93 |

| s = 6 | | | | s = 6 | | |
|---|---|---|---|---|---|---|
| u | $Jo_1(s,u)$ | x | | u | $Jo_1(s,u)$ | x |
| .0000 | .0000 | | | 1.3404 | -.2754 | 5.00 |
| .0205 | .0270 | 2.45 | | 1.4483 | -.4030 | 5.50 |
| .2993 | .3719 | 2.56 | | 1.5661 | -.3938 | 6.12 |
| .4904 | .5432 | 2.75 | | 1.6315 | -.3079 | 6.50 |
| .5504 | .5776 | 2.83 | | 1.6641 | -.2448 | 6.70 |
| .6585 | .6084 | 3.00 | | 1.7111 | -.1355 | 7.00 |
| .8109 | .5690 | 3.30 | | 1.7630 | .0008 | 7.35 |
| .8959 | .4988 | 3.50 | | 1.8524 | .2231 | 8.00 |
| .9609 | .4209 | 3.67 | | 1.9245 | ..3279 | 8.57 |
| 1.0730 | .2416 | 4.00 | | 1.9754 | .3324 | 9.00 |
| 1.1711 | .0507 | 4.33 | | 2.0 | .3098 | 9.22 |
| 1.3172 | -.2340 | 4.90 | | 2.063 | .1858 | 9.80 |

$$Jo_1(s,u) = \sqrt{\frac{\pi}{2}} \ tanh \ u \sum_{k=o}^{\infty} (-1)^k (2k+1) \ Do_{2k+1} \cdot J_{2k+1}(x); \ (cont'd)$$

| | s = 9 | | | s = 15 | |
|---|---|---|---|---|---|
| u | $Jo_1(s,u)$ | x | u | $Jo_1(s,u)$ | x |
| .0000 | .0000 | | .0000 | .0000 | |
| .3974 | .505 | 3.24 | .0602 | .108 | 3.88 |
| .5469 | .570 | 3.46 | .2554 | .404 | 4.00 |
| .5697 | .570 | 3.50 | .4490 | .514 | 4.27 |
| .7953 | .413 | 4.00 | .5616 | .455 | 4.50 |
| .9624 | .123 | 4.50 | .7131 | .230 | 4.90 |
| 1.0987 | -.158 | 5.00 | .7456 | .165 | 5.00 |
| 1.2149 | -.350 | 5.50 | .8087 | .028 | 5.21 |
| 1.2510 | -.387 | 5.67 | .9625 | -.285 | 5.81 |
| 1.3170 | -.415 | 6.00 | 1.0052 | -.346 | 6.00 |
| 1.4083 | -.355 | 6.50 | 1.0996 | -.402 | 6.46 |
| 1.4423 | -.302 | 6.70 | 1.1073 | -.401 | 6.50 |
| 1.4910 | -.202 | 7.00 | 1.1447 | -.383 | 6.70 |
| 1.5668 | -.007 | 7.50 | 1.1979 | -.319 | 7.00 |
| 1.6368 | .175 | 8.00 | 1.3176 | -.045 | 7.75 |
| 1.7627 | .336 | 9.00 | 1.3540 | .055 | 8.00 |
| 1.7883 | .324 | 9.22 | 1.4319 | .244 | 8.57 |
| 1.8526 | .219 | 9.80 | 1.4999 | .328 | 9.11 |
| 1.8738 | .164 | 10.00 | 1.5131 | .332 | 9.22 |
| 1.9248 | .008 | 10.50 | 1.5665 | .297 | 9.68 |
| 1.9733 | -.142 | 11.00 | 1.6019 | .232 | 10.00 |
| 2.0622 | -.289 | 12.00 | 1.6546 | .087 | 10.50 |
| | | | 1.7045 | -.073 | 11.00 |
| | | | 1.7969 | -.279 | 12.00 |
| | | | 1.8811 | -.202 | 13.00 |
| | | | 1.9585 | .060 | 14.00 |
| | | | 2.029 | .246 | 15.00 |

$$Jo_1(s,u) = \sqrt{\frac{\pi}{2}} \tanh u \sum_{k=0}^{\infty} (-1)^k (2k+1) \, Do_{2k+1} \cdot J_{2k+1}(x); \quad \text{(cont'd)}$$

s = 25

| u | $Jo_1(s,u)$ | x |
|---|---|---|
| .0000 | .0000 | |
| .1413 | .277 | 5.05 |
| .3022 | .451 | 5.23 |
| .4436 | .396 | 5.50 |
| .6224 | .062 | 6.00 |
| .6402 | .020 | 6.06 |
| .7141 | -.153 | 6.33 |
| .7564 | -.240 | 6.50 |
| .8671 | -.378 | 7.00 |
| .8792 | -.382 | 7.06 |
| .9624 | -.339 | 7.50 |
| 1.0470 | -.177 | 8.00 |
| 1.0996 | -.035 | 8.34 |
| 1.1929 | .214 | 9.00 |

s = 25

| u | $Jo_1(s,u)$ | x |
|---|---|---|
| 1.2218 | .270 | 9.22 |
| 1.2497 | .307 | 9.44 |
| 1.3170 | .309 | 10.00 |
| 1.3729 | .213 | 10.50 |
| 1.4254 | .060 | 11.00 |
| 1.5221 | -.227 | 12.00 |
| 1.6094 | -.257 | 13.00 |
| 1.6892 | -.029 | 14.00 |
| 1.7628 | .210 | 15.00 |
| 1.8310 | .229 | 16.00 |
| 1.8946 | .027 | 17.00 |
| 1.9542 | -.189 | 18.00 |
| 2.0099 | -.213 | 19.00 |
| 2.0622 | -.035 | 20.00 |

s = 40

| u | $Jo_1(s,u)$ | x |
|---|---|---|
| .0000 | .0000 | |
| .0415 | .099 | 6.33 |
| .2350 | .401 | 6.50 |
| .3429 | .361 | 6.70 |
| .4581 | .146 | 7.00 |
| .5203 | -.011 | 7.20 |
| .6006 | -.208 | 7.50 |
| .7127 | -.358 | 8.00 |
| .8073 | -.294 | 8.50 |
| .9207 | -.010 | 9.20 |
| .9237 | -.001 | 9.22 |
| .9629 | .113 | 9.49 |

s = 40

| u | $Jo_1(s,u)$ | x |
|---|---|---|
| 1.0317 | .270 | 10.00 |
| 1.0937 | .313 | 10.50 |
| 1.1513 | .244 | 11.00 |
| 1.2556 | -.071 | 12.00 |
| 1.3484 | -.275 | 13.00 |
| 1.4323 | -.162 | 14.00 |
| 1.5090 | .113 | 15.00 |
| 1.5797 | .251 | 16.00 |
| 1.6454 | .125 | 17.00 |
| 1.7067 | -.119 | 18.00 |
| 1.7642 | -.231 | 19.00 |
| 1.8185 | -.110 | 20.00 |
| 1.8697 | .111 | 21.00 |
| 1.9184 | .214 | 22.00 |

$$Jo_2(s,u) = -\sqrt{\frac{\pi}{2}}\ \tanh u \sum_{k=o}^{\infty} (-1)^k (2k)\, Do_{2k} \cdot J_{2k}(x); \quad x = \sqrt{s}\ \cosh u$$

| | s = 1 | | | s = 3 | |
|---|---|---|---|---|---|
| u | $Jo_2(s,u)$ | x | u | $Jo_2(s,u)$ | x |
| .0000 | .0000 | | .0000 | .0000 | |
| .4436 | .0744 | 1.10 | .1441 | .0605 | 1.75 |
| .6224 | .1153 | 1.20 | .2795 | .1200 | 1.80 |
| .7564 | .1531 | 1.30 | .3670 | .1609 | 1.85 |
| .8671 | .1900 | 1.40 | .5493 | .2544 | 2.00 |
| 1.0470 | .2629 | 1.60 | .9100 | .4678 | 2.50 |
| 1.1929 | .3338 | 1.80 | 1.0371 | .5384 | 2.75 |
| 1.3169 | .4010 | 2.00 | 1.1462 | .5847 | 3.00 |
| 1.4254 | .4623 | 2.20 | 1.1861 | .5961 | 3.10 |
| 1.5221 | .5153 | 2.40 | 1.3289 | .5996 | 3.50 |
| 1.5668 | .5381 | 2.50 | 1.4796 | .5130 | 4.00 |
| 1.6892 | .5886 | 2.80 | 1.6086 | .3441 | 4.50 |
| 1.7628 | .6059 | 3.00 | 1.7218 | .1282 | 5.00 |
| 1.8946 | .5967 | 3.40 | 1.8228 | -.0906 | 5.50 |
| 2.0099 | .5294 | 3.80 | 1.9974 | -.3733 | 6.50 |

| | s = 6 | | | s = 9 | |
|---|---|---|---|---|---|
| u | $Jo_2(s,u)$ | x | u | $Jo_2(s,u)$ | x |
| .0000 | .0000 | | .0000 | .0000 | |
| .0205 | .0152 | 2.45 | .3045 | .2996 | 3.14 |
| .4904 | .3696 | 2.75 | .5697 | .5157 | 3.50 |
| .6585 | .4882 | 3.00 | .7954 | .5811 | 4.00 |
| .8959 | .5932 | 3.50 | 1.0987 | .2925 | 5.00 |
| 1.0730 | .5599 | 4.00 | 1.2149 | .0558 | 5.50 |
| 1.3404 | .2070 | 5.00 | 1.3170 | -.1654 | 6.00 |
| 1.4483 | -.0231 | 5.50 | 1.4083 | -.3237 | 6.50 |
| 1.6315 | -.3534 | 6.50 | 1.4910 | -.3895 | 7.00 |
| 1.7111 | -.3914 | 7.00 | 1.6368 | -.2401 | 8.00 |
| 1.8524 | -.2065 | 8.00 | 1.7627 | .0985 | 9.00 |
| 2.0 | .193 | 9.22 | 1.7883 | .1659 | 9.22 |
| | | | 1.8738 | .3109 | 10.00 |
| | | | 1.9733 | .2275 | 11.00 |

$$Jo_2(s,u) = -\sqrt{\frac{\pi}{2}} \tanh u \sum_{k=o}^{\infty} (-1)^k (2k) Do_{2k} \cdot J_{2k}(x); \quad \text{(cont'd)}$$

| s = 15 | | | s = 15 | | |
|---|---|---|---|---|---|
| u | $Jo_2(s,u)$ | x | u | $Jo_2(s,u)$ | x |
| .0000 | .0000 | | 1.2605 | -.387 | 7.38 |
| .0602 | .083 | 3.88 | 1.3540 | -.304 | 8.00 |
| .2554 | .334 | 4.00 | 1.5131 | .101 | 9.22 |
| .5013 | .536 | 4.37 | 1.6019 | .290 | 10.00 |
| .5616 | .550 | 4.50 | 1.6505 | .317 | 10.46 |
| .7456 | .460 | 5.00 | 1.7045 | .261 | 11.00 |
| 1.0052 | -.011 | 6.00 | 1.7969 | -.003 | 12.00 |
| 1.1073 | -.228 | 6.50 | 1.8811 | -.241 | 13.00 |
| 1.1979 | -.359 | 7.00 | 1.9585 | -.241 | 14.00 |

| s = 25 | | | s = 40 | | |
|---|---|---|---|---|---|
| u | $Jo_2(s,u)$ | x | u | $Jo_2(s,u)$ | x |
| .0000 | .0000 | | .0000 | .0000 | |
| .3022 | .445 | 5.23 | .0415 | .089 | 6.33 |
| .4436 | .497 | 5.50 | .2350 | .405 | 6.50 |
| .6224 | .331 | 6.00 | .4581 | .324 | 7.00 |
| .7141 | .151 | 6.33 | .6006 | .014 | 7.50 |
| .7564 | .055 | 6.50 | .7127 | -.245 | 8.00 |
| .8671 | -.196 | 7.00 | .8073 | -.360 | 8.50 |
| 1.0008 | -.374 | 7.72 | .9237 | -.267 | 9.22 |
| 1.0470 | -.373 | 8.00 | 1.0317 | .021 | 10.00 |
| 1.1929 | -.121 | 9.00 | 1.1513 | .294 | 11.00 |
| 1.2218 | -.039 | 9.22 | 1.2556 | .216 | 12.00 |
| 1.3169 | .219 | 10.00 | 1.3484 | -.079 | 13.00 |
| 1.4254 | .303 | 11.00 | 1.4323 | -.267 | 14.00 |
| 1.5221 | .085 | 12.00 | 1.5090 | -.174 | 15.00 |
| 1.6094 | -.193 | 13.00 | 1.5797 | .084 | 16.00 |
| 1.6892 | -.266 | 14.00 | 1.6454 | .242 | 17.00 |
| 1.7628 | -.083 | 15.00 | 1.7067 | .157 | 18.00 |
| 1.8309 | .164 | 16.00 | 1.8185 | -.221 | 20.00 |
| 1.8946 | .242 | 17.00 | 1.8697 | -.152 | 21.00 |
| 1.9542 | .091 | 18.00 | 1.918 | .057 | 22.00 |

## TABLE III

### Radial Mathieu Functions of the Second Kind

$$Ne_o(s,u) = \sqrt{\frac{\pi}{2}} \sum_{k=0}^{\infty} (-1)^k De_{2k} \left[ \frac{Y_k(y) \cdot J_k(x)}{De_o} \right], \quad y = \frac{\sqrt{s}}{2} e^u, \quad x = \frac{\sqrt{s}}{2} e^{-u}$$

s = 1

| u | $Ne_o(s,u)$ | y | x |
|---|---|---|---|
| 0 | -.57872 | | |
| .182 | -.410 | .60 | .417 |
| .470 | -.130 | .80 | .312 |
| .588 | -.012 | .90 | .278 |
| .693 | +.094 | 1.00 | .250 |
| .876 | +.273 | 1.20 | .208 |
| .956 | +.348 | 1.30 | .192 |
| 1.224 | +.560 | 1.70 | .147 |
| 1.281 | +.593 | 1.80 | .139 |
| 1.435 | +.647 | 2.10 | .119 |
| 1.526 | +.648 | 2.30 | .109 |
| 1.609 | +.624 | 2.50 | .100 |
| 1.723 | +.547 | 2.80 | .089 |
| 1.792 | +.474 | 3.00 | .083 |
| 1.974 | +.187 | 3.60 | .069 |

s = 3

| u | $Ne_o(s,u)$ | y | x |
|---|---|---|---|
| 0 | -.18776 | | |
| .144 | -.028 | 1.00 | .75 |
| .480 | +.337 | 1.40 | .54 |
| .614 | +.462 | 1.60 | .469 |
| .732 | +.552 | 1.80 | .417 |
| .886 | +.625 | 2.10 | .357 |
| 1.019 | +.631 | 2.40 | .312 |
| 1.242 | +.483 | 3.00 | .250 |
| 1.530 | -.004 | 4.00 | .188 |
| 1.753 | -.380 | 5.00 | .150 |
| 1.936 | -.365 | 6.00 | .125 |

s = 6

| u | $Ne_o(s,u)$ | y | x |
|---|---|---|---|
| 0 | -.05169 | | |
| .060 | +.030 | 1.30 | 1.15 |
| .203 | +.217 | 1.50 | 1.00 |
| .490 | +.521 | 2.00 | .75 |
| .714 | +.601 | 2.50 | .60 |
| .827 | +.560 | 2.80 | .54 |
| .896 | +.504 | 3.00 | .500 |
| .991 | +.387 | 3.30 | .455 |
| 1.106 | +.193 | 3.70 | .405 |
| 1.184 | +.037 | 4.00 | .375 |
| 1.407 | -.361 | 5.00 | .300 |
| 1.589 | -.372 | 6.00 | .250 |
| 1.743 | -.057 | 7.00 | .214 |
| 1.877 | +.266 | 8.00 | .188 |
| 1.994 | +.318 | 9.00 | .167 |

s = 9

| u | $Ne_o(s,u)$ | y | x |
|---|---|---|---|
| 0 | -.01781 | | |
| .064 | +.080 | 1.60 | 1.41 |
| .182 | +.254 | 1.80 | 1.25 |
| .470 | +.543 | 2.40 | .94 |
| .550 | +.566 | 2.60 | .87 |
| .693 | +.522 | 3.00 | .75 |
| .903 | +.245 | 3.70 | .61 |
| .981 | +.090 | 4.00 | .56 |
| 1.099 | -.153 | 4.50 | .500 |
| 1.204 | -.332 | 5.00 | .450 |
| 1.386 | -.381 | 6.00 | .375 |
| 1.540 | -.080 | 7.00 | .321 |
| 1.674 | +.252 | 8.00 | .281 |
| 1.792 | +.321 | 9.00 | .250 |
| 1.897 | +.097 | 10.00 | .225 |
| 1.992 | -.193 | 11.00 | .205 |

$$Ne_0(s,u) = \sqrt{\frac{\pi}{2}} \sum_{k=o}^{\infty} (-1)^k De_{2k}\left[\frac{Y_k(y) \cdot J_k(x)}{De_o}\right], \quad \text{(cont'd)}$$

s = 15

| u | $Ne_0(s,u)$ | y | x |
|---|---|---|---|
| 0 | -.00314 | | |
| .032 | +.057 | 2.00 | 1.88 |
| .255 | +.403 | 2.50 | 1.50 |
| .438 | +.514 | 3.00 | 1.25 |
| .592 | +.423 | 3.50 | 1.07 |
| .725 | +.207 | 4.00 | .94 |
| .821 | +.002 | 4.40 | .85 |
| 1.131 | -.392 | 6.00 | .62 |
| 1.285 | -.134 | 7.00 | .54 |
| 1.419 | +.216 | 8.00 | .469 |
| 1.536 | +.326 | 9.00 | .417 |
| 1.642 | +.127 | 10.00 | .375 |
| 1.737 | -.169 | 11.00 | .341 |
| 1.824 | -.286 | 12.00 | .312 |

s = 25

| u | $Ne_0(s,u)$ | y | x |
|---|---|---|---|
| 0 | -.00033 | | |
| .182 | +.337 | 3.00 | 2.08 |
| .336 | +.457 | 3.50 | 1.79 |
| .365 | +.453 | 3.60 | 1.74 |
| .470 | +.362 | 4.00 | 1.56 |
| .652 | -.010 | 4.80 | 1.30 |
| .693 | -.105 | 5.00 | 1.25 |
| .876 | -.381 | 6.00 | 1.04 |
| 1.030 | -.218 | 7.00 | .89 |
| 1.163 | +.142 | 8.00 | .78 |
| 1.281 | +.324 | 9.00 | .69 |
| 1.386 | +.178 | 10.00 | .62 |
| 1.482 | -.124 | 11.00 | .57 |
| 1.569 | -.285 | 12.00 | .52 |

$$Ne_1(s,u) = \sqrt{\frac{\pi}{2}} \sum_{k=o}^{\infty} (-1)^k \frac{De_{2k+1}}{De_1}\left[Y_{k+1}(y) \cdot J_k(x) + Y_k(y) \cdot J_{k+1}(x)\right],$$

$$y = \frac{\sqrt{s}}{2} e^u, \quad x = \frac{\sqrt{s}}{2} e^{-u}$$

s = 1

| u | $Ne_1(s,u)$ | y | x |
|---|---|---|---|
| 0 | -1.92135 | | |
| .182 | -1.624 | .60 | .417 |
| .470 | -1.228 | .80 | .312 |
| .588 | -1.083 | .90 | .278 |
| .693 | -.959 | 1.00 | .250 |
| .876 | -.746 | 1.20 | .208 |
| .956 | -.652 | 1.30 | .192 |
| 1.224 | -.315 | 1.70 | .147 |
| 1.435 | -.027 | 2.10 | .119 |
| 1.609 | +.213 | 2.50 | .100 |
| 1.723 | +.354 | 2.80 | .089 |
| 1.792 | +.425 | 3.00 | .083 |
| 1.974 | +.526 | 3.60 | .069 |

s = 3

| u | $Ne_1(s,u)$ | y | x |
|---|---|---|---|
| 0 | -1.06409 | | |
| .144 | -.890 | 1.00 | .75 |
| .480 | -.484 | 1.40 | .54 |
| .614 | -.318 | 1.60 | .469 |
| .732 | -.165 | 1.80 | .417 |
| 1.019 | +.213 | 2.40 | .312 |
| 1.099 | +.312 | 2.60 | .288 |
| *1.146 | +.364 | | |
| *1.329 | +.514 | | |
| 1.452 | +.532 | 3.70 | .203 |
| *1.559 | +.471 | | |
| *1.722 | +.224 | | |
| 1.753 | +.159 | 5.00 | .150 |
| *1.914 | -.198 | | |

* - See Equations (18) and (19)

36

$$Ne_1(s,u) = \sqrt{\frac{\pi}{2}} \sum_{k=0}^{\infty} (-1)^k \frac{De_{2k+1}}{De_1} \left[ Y_{k+1}(y) \cdot J_k(x) + Y_k(y) \cdot J_{k+1}(x) \right], \text{ (cont'd)}$$

s = 6

| u | $Ne_1(s,u)$ | y | x |
|---|---|---|---|
| 0 | -.55600 | | |
| .060 | -.489 | 1.30 | 1.15 |
| .203 | -.317 | 1.50 | 1.00 |
| .490 | +.044 | 2.00 | .75 |
| .714 | +.316 | 2.50 | .60 |
| .827 | +.433 | 2.80 | .54 |
| .991 | +.536 | 3.30 | .455 |
| 1.106 | +.536 | 3.70 | .405 |
| *1.163 | +.507 | | |
| *1.317 | +.316 | | |
| *1.340 | +.274 | | |
| 1.407 | +.159 | 5.00 | .300 |
| *1.544 | -.166 | | |
| *1.711 | -.384 | | |
| *1.852 | -.235 | | |
| *1.975 | +.099 | | |

s = 9

| u | $Ne_1(s,u)$ | y | x |
|---|---|---|---|
| 0 | -.29671 | | |
| .182 | -.070 | 1.80 | 1.25 |
| .470 | +.289 | 2.40 | .94 |
| .693 | +.500 | 3.00 | .75 |
| .981 | +.487 | 4.00 | .56 |
| 1.099 | +.335 | 4.50 | .500 |
| *1.317 | -.123 | | |
| *1.491 | -.380 | | |
| *1.637 | -.258 | | |
| *1.763 | +.077 | | |
| *1.874 | +.304 | | |
| *1.973 | +.237 | | |
| *2.062 | -.035 | | |

s = 15

| u | $Ne_1(s,u)$ | y | x |
|---|---|---|---|
| 0 | -.08494 | | |
| .032 | -.039 | 2.00 | 1.88 |
| .255 | +.274 | 2.50 | 1.50 |
| .438 | +.470 | 3.00 | 1.25 |
| .592 | +.539 | 3.50 | 1.07 |
| .725 | +.487 | 4.00 | .94 |
| .949 | +.136 | 5.00 | .75 |
| 1.131 | -.256 | 6.00 | .62 |
| 1.285 | -.380 | 7.00 | .54 |
| 1.419 | ~~-.137~~ -.174 | 8.00 | .469 |
| 1.536 | ~~+.181~~ .154 | 9.00 | .417 |
| 1.642 | +.317 | 10.00 | .375 |
| 1.737 | ~~.161~~ .122 | 11.00 | .341 |
| 1.824 | ~~-.119~~ -.087 | 12.00 | .312 |

s = 25

| u | $Ne_1(s,u)$ | y | x |
|---|---|---|---|
| 0 | -.01251 | | |
| .182 | +.296 | 3.00 | 2.08 |
| .365 | +.481 | 3.60 | 1.74 |
| .470 | +.489 | 4.00 | 1.56 |
| .693 | +.201 | 5.00 | 1.25 |
| .876 | -.209 | 6.00 | 1.04 |
| 1.030 | -.378 | 7.00 | .89 |
| 1.163 | -.201 | 8.00 | .78 |
| 1.281 | +.129 | 9.00 | .69 |
| 1.386 | +.312 | 10.00 | .62 |
| 1.482 | +.201 | 11.00 | .57 |
| 1.569 | -.070 | 12.00 | .52 |

* - See Equations (18) and (19)

$$Ne_2(s,u) = -\sqrt{\frac{\pi}{2}} \sum_{k=0}^{\infty} \left[(-1)^k \cdot De_{2k} \cdot Y_k(y) \cdot J_k(x) / De_o\right], \quad y = \frac{\sqrt{s}}{2} e^u, \quad x = \frac{\sqrt{s}}{2} e^{-u}$$

s = 1

| u | $Ne_2(s,u)$ | y | x |
|---|---|---|---|
| 0 | -6.74259 | | |
| .182 | -4.9 | .60 | .417 |
| .336 | -3.7 | .70 | .357 |
| .470 | -2.9 | .80 | .312 |
| .588 | -2.4 | .90 | .278 |
| .693 | -2.1 | 1.00 | .250 |
| .788 | -1.8 | 1.10 | .227 |
| .876 | -1.6 | 1.20 | .208 |
| .956 | -1.4 | 1.30 | .192 |
| 1.030 | -1.3 | 1.40 | .179 |
| 1.224 | -1.0 | 1.70 | .147 |
| 1.435 | -0.7 | 2.10 | .119 |
| 1.609 | -0.5 | 2.50 | .100 |
| 1.723 | -0.3 | 2.80 | .089 |
| 1.792 | -0.2 | 3.00 | .083 |
| 1.974 | +0.1 | 3.60 | .069 |

s = 3

| u | $Ne_2(s,u)$ | y | x |
|---|---|---|---|
| 0 | -2.57520 | | |
| .144 | -2.07 | 1.00 | .75 |
| .239 | -1.79 | 1.10 | .68 |
| .480 | -1.29 | 1.40 | .54 |
| .614 | -1.07 | 1.60 | .469 |
| .732 | -0.90 | 1.80 | .417 |
| 1.019 | -0.52 | 2.40 | .312 |
| 1.242 | -0.18 | 3.00 | .250 |
| 1.530 | +0.28 | 4.00 | .188 |
| 1.753 | +0.47 | 5.00 | .150 |
| 1.936 | +0.28 | 6.00 | .125 |

s = 6

| u | $Ne_2(s,u)$ | y | x |
|---|---|---|---|
| 0 | -1.51795 | | |
| .060 | -1.40 | 1.30 | 1.15 |
| .203 | -1.14 | 1.50 | 1.00 |
| .490 | -0.72 | 2.00 | .75 |
| .714 | -0.41 | 2.50 | .60 |
| .827 | -0.24 | 2.80 | .54 |
| .896 | -0.13 | 3.00 | .500 |
| .991 | +0.02 | 3.30 | .455 |
| 1.106 | +0.21 | 3.70 | .405 |
| 1.184 | +0.32 | 4.00 | .375 |
| 1.345 | +0.46 | 4.70 | .319 |
| 1.407 | +0.47 | 5.00 | .300 |
| 1.589 | +0.27 | 6.00 | .250 |
| 1.743 | -0.09 | 7.00 | .214 |
| 1.877 | -0.34 | 8.00 | .188 |
| 1.994 | -0.28 | 9.00 | .167 |

s = 9

| u | $Ne_2(s,u)$ | y | x |
|---|---|---|---|
| 0 | -1.08259 | | |
| .064 | -0.98 | 1.60 | 1.41 |
| .182 | -0.81 | 1.80 | 1.25 |
| .470 | -0.41 | 2.40 | .94 |
| .550 | -.29 | 2.60 | .87 |
| .693 | -.07 | 3.00 | .75 |
| .903 | +0.25 | 3.70 | .61 |
| .981 | +0.35 | 4.00 | .56 |
| 1.099 | +0.45 | 4.50 | .500 |
| 1.204 | +0.47 | 5.00 | .450 |
| 1.386 | +0.25 | 6.00 | .375 |
| 1.540 | -0.11 | 7.00 | .321 |
| 1.674 | -0.34 | 8.00 | .281 |
| 1.792 | -0.27 | 9.00 | .250 |
| 1.897 | +0.01 | 10.00 | .225 |
| 1.992 | +0.26 | 11.00 | .205 |

$$Ne_2(s,u) = \sqrt{\frac{\pi}{2}} \sum_{k=o}^{\infty} \left[ (-1)^k \cdot De_{2k} \cdot Y_k(y) \cdot J_k(x)/De_o \right], \quad \text{(cont'd)}$$

| | s = 15 | | | | s = 25 | | |
|---|---|---|---|---|---|---|---|
| u | $Ne_2(s,u)$ | y | x | u | $Ne_2(s,u)$ | y | x |
| 0 | -0.57338 | | | 0 | -0.17018 | | |
| .032 | -0.53 | 2.00 | 1.88 | .182 | +0.09 | 3.00 | 2.08 |
| .255 | -0.23 | 2.50 | 1.50 | .470 | +0.44 | 4.00 | 1.56 |
| .438 | +0.03 | 3.00 | 1.25 | .652 | +0.49 | 4.80 | 1.30 |
| .592 | +0.25 | 3.50 | 1.07 | .693 | +0.47 | 5.00 | 1.25 |
| .725 | +0.41 | 4.00 | .94 | .876 | +0.19 | 6.00 | 1.04 |
| .821 | +0.47 | 4.40 | .85 | 1.030 | -0.17 | 7.00 | .89 |
| .949 | +0.47 | 5.00 | .75 | 1.163 | -0.36 | 8.00 | .78 |
| 1.131 | +0.22 | 6.00 | .62 | 1.281 | -0.24 | 9.00 | .69 |
| 1.285 | -0.16 | 7.00 | .54 | 1.386 | +0.05 | 10.00 | .62 |
| 1.419 | -0.37 | 8.00 | .469 | 1.482 | +0.33 | 11.00 | .57 |
| 1.536 | -0.26 | 9.00 | .417 | 1.569 | +0.33 | 12.00 | .52 |
| 1.642 | +0.03 | 10.00 | .375 | | | | |
| 1.737 | +0.27 | 11.00 | .341 | | | | |
| 1.824 | +0.26 | 12.00 | .312 | | | | |

$$No_1(s,u) = \sqrt{\frac{\pi}{2}} \sum_{k=o}^{\infty} (-1)^k \cdot \frac{Do_{2k+1}}{Do_1} \cdot \left[ Y_{k+1}(y) \cdot J_k(x) - Y_k(y) \cdot J_{k+1}(x) \right],$$

$$y = \frac{\sqrt{s}}{2} e^u, \quad x = \frac{\sqrt{s}}{2} e^{-u}$$

| | s = 1 | | | | s = 3 | | |
|---|---|---|---|---|---|---|---|
| u | $No_1(s,u)$ | y | x | u | $No_1(s,u)$ | y | x |
| 0 | -1.64460 | | | 0 | -1.00233 | | |
| .182 | -1.462 | .60 | .417 | .144 | -.942 | 1.00 | .75 |
| .470 | -1.193 | .80 | .312 | .480 | -.695 | 1.40 | .54 |
| .588 | -1.084 | .90 | .278 | .614 | -.552 | 1.60 | .469 |
| .693 | -.985 | 1.00 | .250 | .732 | -.406 | 1.80 | .417 |
| .876 | -.805 | 1.20 | .208 | 1.019 | +.018 | 2.40 | .312 |
| .956 | -.719 | 1.30 | .192 | 1.099 | +.139 | 2.60 | .288 |
| 1.224 | -.399 | 1.70 | .147 | 1.242 | +.340 | 3.00 | .250 |
| 1.435 | -.105 | 2.10 | .119 | *1.329 | +.399 | | |
| 1.609 | +.151 | 2.50 | .100 | 1.452 | +.504 | 3.70 | .203 |
| 1.723 | +.305 | 2.80 | .089 | *1.559 | +.484 | | |
| 1.792 | +.387 | 3.00 | .083 | *1.722 | +.278 | | |
| 1.974 | +.513 | 3.60 | .069 | *1.914 | -.149 | | |

\* - See Equations (18) and (19)

$$No_1(s,u) = \sqrt{\frac{\pi}{2}} \sum_{k=0}^{\infty} (-1)^k \cdot \frac{Do_{2k+1}}{Do_1} \left[ Y_{k+1}(y) \cdot J_k(x) - Y_k(y) \cdot J_{k+1}(x) \right], \text{ (cont'd)}$$

s = 6

| u | $No_1(s,u)$ | y | x |
|---|---|---|---|
| 0 | -.75888 | | |
| .060 | -.748 | 1.30 | 1.15 |
| .203 | -.680 | 1.50 | 1.00 |
| .490 | -.381 | 2.00 | .75 |
| .714 | -.035 | 2.50 | .60 |
| .827 | +.150 | 2.80 | .54 |
| .991 | +.384 | 3.30 | .455 |
| 1.106 | +.478 | 3.70 | .405 |
| *1.163 | +.493 | | |
| *1.317 | +.397 | | |
| *1.340 | +.365 | | |
| *1.544 | -.064 | | |
| *1.711 | -.357 | | |
| *1.852 | -.278 | | |
| *1.975 | +.045 | | |

s = 9

| u | $No_1(s,u)$ | y | x |
|---|---|---|---|
| 0 | -.65550 | | |
| .182 | -.580 | 1.80 | 1.25 |
| .470 | -.221 | 2.40 | .94 |
| .693 | +.165 | 3.00 | .75 |
| .903 | +.440 | 3.70 | .61 |
| .981 | +.477 | 4.00 | .56 |
| 1.099 | +.432 | 4.50 | .500 |
| *1.317 | +.032 | | |
| *1.491 | -.325 | | |
| 1.540 | -.366 | 7.00 | .321 |
| *1.637 | -.312 | | |
| *1.763 | -.003 | | |
| *1.874 | +.273 | | |
| *1.973 | +.268 | | |
| *2.062 | +.016 | | |

s = 15

| u | $No_1(s,u)$ | y | x |
|---|---|---|---|
| 0 | -.55425 | | |
| .032 | -.551 | 2.00 | 1.88 |
| .255 | -.356 | 2.50 | 1.50 |
| .438 | -.035 | 3.00 | 1.25 |
| ..592 | +.253 | 3.50 | 1.07 |
| .725 | +.421 | 4.00 | .94 |
| .821 | +.452 | 4.40 | .85 |
| .949 | +.340 | 5.00 | .75 |
| 1.131 | -.061 | 6.00 | .62 |
| 1.285 | -.345 | 7.00 | .54 |
| 1.419 | +.255 .272 | 8.00 | .469 |
| 1.536 | +.055 .042 | 9.00 | .417 |
| 1.642 | +.286 | 10.00 | .375 |
| 1.737 | +.232 .246 | 11.00 | .341 |
| 1.824 | -.029 .014 | 12.00 | .312 |

s = 25

| u | $No_1(s,u)$ | y | x |
|---|---|---|---|
| 0 | -.47546 | | |
| .182 | -.322 | 3.00 | 2.08 |
| .365 | +.040 | 3.60 | 1.74 |
| .470 | +.256 | 4.00 | 1.56 |
| .652 | +.419 | 4.80 | 1.30 |
| .693 | +.400 | 5.00 | 1.25 |
| .876 | +.066 | 6.00 | 1.04 |
| 1.030 | -.290 | 7.00 | .89 |
| 1.099 | -.372 | 7.50 | .83 |
| 1.163 | -.312 | 8.00 | .78 |
| 1.281 | -.027 | 9.00 | .69 |
| 1.386 | +.254 | 10.00 | .62 |
| 1.482 | +.273 | 11.00 | .57 |
| 1.569 | +.032 | 12.00 | .52 |

* - See Equations (18) and (19)

$$No_2(s,u) = \sqrt{\frac{\pi}{2}} \sum_{k=1}^{\infty} (-1)^{k+1} \frac{Do_{2k}}{Do_2} \left[ Y_{k+1}(y) \cdot J_{k-1}(x) - Y_{k-1}(y) \cdot J_{k+1}(x) \right],$$

$$y = \frac{\sqrt{s}}{2} e^u, \quad x = \frac{\sqrt{s}}{2} e^{-u}$$

s = 1

| u | $No_2(s,u)$ | y | x |
|---|---|---|---|
| 0 | -6.65210 | | |
| .182 | -4.791 | .60 | .417 |
| .336 | -3.663 | .70 | .357 |
| .470 | -2.928 | .80 | .312 |
| .588 | -2.420 | .90 | .278 |
| .693 | -2.056 | 1.00 | .250 |
| .788 | -1.785 | 1.10 | .227 |
| .876 | -1.577 | 1.20 | .208 |
| .956 | -1.413 | 1.30 | .192 |
| 1.030 | -1.278 | 1.40 | .179 |
| 1.224 | -.985 | 1.70 | .147 |
| 1.281 | -.909 | 1.80 | .139 |
| 1.435 | -.711 | 2.10 | .119 |
| 1.526 | -.593 | 2.30 | .109 |
| 1.609 | -.479 | 2.50 | .100 |
| 1.723 | -.311 | 2.80 | .089 |
| 1.792 | -.202 | 3.00 | .083 |
| 1.974 | +.104 | 3.60 | .069 |

s = 3

| u | $No_2(s,u)$ | y | x |
|---|---|---|---|
| 0 | -2.40248 | | |
| .144 | -1.958 | 1.00 | .75 |
| .239 | -1.719 | 1.10 | .68 |
| .480 | -1.255 | 1.40 | .54 |
| .614 | -1.060 | 1.60 | .469 |
| .732 | -.906 | 1.80 | .417 |
| 1.019 | -.539 | 2.40 | .312 |
| 1.242 | -.207 | 3.00 | .250 |
| 1.368 | +.003 | 3.40 | .221 |
| 1.530 | +.267 | 4.00 | .188 |
| 1.753 | +.460 | 5.00 | .150 |
| 1.936 | +.289 | 6.00 | .125 |

s = 6

| u | $No_2(s,u)$ | y | x |
|---|---|---|---|
| 0 | -1.34633 | | |
| .060 | -1.270 | 1.30 | 1.15 |
| .203 | -1.096 | 1.50 | 1.00 |
| .490 | -.774 | 2.00 | .75 |
| .714 | -.497 | 2.50 | .60 |
| .827 | -.333 | 2.80 | .54 |
| .896 | -.224 | 3.00 | .500 |
| .991 | -.064 | 3.30 | .455 |
| 1.106 | +.131 | 3.70 | .405 |
| 1.184 | +.256 | 4.00 | .375 |
| 1.407 | +.458 | 5.00 | .300 |
| 1.589 | +.293 | 6.00 | .250 |
| 1.743 | -.070 | 7.00 | .214 |
| 1.877 | -.328 | 8.00 | .188 |
| 1.994 | -.286 | 9.00 | .167 |

s = 9

| u | $No_2(s,u)$ | y | x |
|---|---|---|---|
| 0 | -.99820 | | |
| .064 | -.952 | 1.60 | 1.41 |
| .182 | -.862 | 1.80 | 1.25 |
| .470 | -.570 | 2.40 | .94 |
| .550 | -.466 | 2.60 | .87 |
| .693 | -.250 | 3.00 | .75 |
| .903 | +.110 | 3.70 | .61 |
| .981 | +.239 | 4.00 | .56 |
| 1.099 | +.393 | 4.50 | .500 |
| 1.204 | +.455 | 5.00 | .450 |
| 1.386 | +.298 | 6.00 | .375 |
| 1.540 | -.064 | 7.00 | .321 |
| 1.674 | -.325 | 8.00 | .281 |
| 1.735 | -.346 | 8.50 | .265 |
| 1.792 | -.288 | 9.00 | .250 |
| 1.897 | -.014 | 10.00 | .225 |
| 1.992 | +.245 | 11.00 | .205 |

$$No_2(s,u) = \sqrt{\frac{\pi}{2}} \sum_{k=1}^{\infty} (-1)^{k+1} \frac{Do_{2k}}{Do_2} \left[ Y_{k+1}(y) \cdot J_{k-1}(x) - Y_{k-1}(y) \cdot J_{k+1}(x) \right] , \text{(cont'd)}$$

| | s = 15 | | | | s = 25 | | |
|---|---|---|---|---|---|---|---|
| u | $No_2(s,u)$ | y | x | u | $No_2(s,u)$ | y | x |
| 0 | -.72398 | | | 0 | -.56114 | | |
| .032 | -.717 | 2.00 | 1.88 | .182 | -.450 | 3.00 | 2.08 |
| .255 | -.564 | 2.50 | 1.50 | .470 | +.047 | 4.00 | 1.56 |
| .438 | -.324 | 3.00 | 1.25 | .693 | +.400 | 5.00 | 1.25 |
| .592 | -.058 | 3.50 | 1.07 | .876 | +.348 | 6.00 | 1.04 |
| .725 | +.183 | 4.00 | .94 | 1.030 | +.008 | 7.00 | .89 |
| .949 | +.442 | 5.00 | .75 | 1.163 | -.293 | 8.00 | .78 |
| 1.131 | +.314 | 6.00 | .62 | 1.281 | -.307 | 9.00 | .69 |
| 1.285 | -.043 | 7.00 | .54 | 1.386 | -.054 | 10.00 | .62 |
| 1.419 | -.317 | 8.00 | .469 | 1.482 | +.217 | 11.00 | .57 |
| 1.536 | -.294 | 9.00 | .417 | 1.569 | +.274 | 12.00 | .52 |
| 1.642 | -.026 | 10.00 | .375 | | | | |
| 1.737 | +.239 | 11.00 | .341 | | | | |
| 1.824 | +.275 | 12.00 | .312 | | | | |

42

# TABLE IV
## Radial Mathieu Functions of the Third Kind

$$He_o^{(1)}(-|s|,u) = -i\sqrt{\frac{2}{\pi}}\left(\frac{1}{De_o}\right)\sum_{m=o}^{\infty} De_{2m}\cdot I_m(x)\cdot K_m(y) ;$$

$$y = \frac{\sqrt{|s|}}{2}\,e^u, \quad x = \frac{\sqrt{|s|}}{2}\,e^{-u}$$

| u | $He_o^{(1)}(-\lvert s\rvert,u)$ | y | x | u | $He_o^{(1)}(-\lvert s\rvert,u)$ | y | x |
|---|---|---|---|---|---|---|---|
| | **s = 1** | | | | **s = 3** | | |
| .000 | -.7425i | .500 | .500 | .000 | -.3891i | .866 | .866 |
| .182 | -.6205i | .60 | .42 | .038 | -.3690i | .90 | .83 |
| .470 | -.4486i | .80 | .31 | .144 | -.3180i | 1.00 | .75 |
| .693 | -.3337i | 1.00 | .25 | .326 | -.2399i | 1.20 | .62 |
| .956 | -.2205i | 1.30 | .19 | .480 | -.1838i | 1.40 | .54 |
| 1.224 | -.1312i | 1.70 | .15 | .614 | -.1421i | 1.60 | .47 |
| 1.435 | -.0800i | 2.10 | .12 | .732 | -.1106i | 1.80 | .42 |
| 1.723 | -.0305i | 2.80 | .09 | .837 | -.0866i | 2.00 | .38 |
| 2.197 | -.0045i | 4.50 | .01 | 1.019 | -.0537i | 2.40 | .31 |
| | | | | 1.099 | -.0424i | 2.60 | .29 |
| | | | | 1.466 | -.0114i | 3.75 | .20 |
| | **s = 6** | | | | **s = 9** | | |
| .000 | -.2028i | 1.225 | 1.225 | .000 | -.1200i | 1.500 | 1.500 |
| .060 | -.1832i | 1.30 | 1.15 | .064 | -.1060i | 1.60 | 1.41 |
| .203 | -.1413i | 1.50 | 1.00 | .182 | -.0834i | 1.80 | 1.25 |
| .490 | -.0769i | 2.00 | .75 | .288 | -.0661i | 2.00 | 1.12 |
| .714 | -.0430i | 2.50 | .60 | .470 | -.0420i | 2.40 | .94 |
| .827 | -.0306i | 2.80 | .54 | .550 | -.0336i | 2.60 | .87 |
| .896 | -.0244i | 3.00 | .50 | .693 | -.0217i | 3.00 | .75 |
| .991 | -.0175i | 3.30 | .45 | .903 | -.0101i | 3.70 | .61 |
| 1.106 | -.0112i | 3.70 | .40 | .981 | -.0073i | 4.00 | .56 |
| 1.407 | -.0027i | 5.00 | .30 | 1.099 | -.0043i | 4.50 | .50 |
| | **s = 15** | | | | **s = 25** | | |
| .000 | -.0508i | 1.936 | 1.936 | .000 | -.0167i | 2.500 | 2.500 |
| .032 | -.0475i | 2.00 | 1.875 | .182 | -.0102i | 3.00 | 2.08 |
| .255 | -.0278i | 2.50 | 1.50 | .470 | -.0040i | 4.00 | 1.56 |
| .438 | -.0165i | 3.00 | 1.25 | .693 | -.0015i | 5.00 | 1.25 |
| .725 | -.0059i | 4.00 | .94 | | | | |

$$He_1^{(1)}(-|s|,u) = -\sqrt{\frac{2}{\pi}}\left(\frac{1}{Do_1}\right)\sum_{m=o}^{\infty}Do_{2m+1}\left[I_m(x)\cdot K_{m+1}(y) - I_{m+1}(x)\cdot K_m(y)\right];$$

$$y = \frac{\sqrt{|s|}}{2}\,e^u, \quad x = \frac{\sqrt{|s|}}{2}\,e^{-u}$$

s = 1

| u | $He_1^{(1)}(-|s|,u)$ | y | x |
|---|---|---|---|
| .000 | -1.1700 | .500 | .500 |
| .182 | - .9270 | .60 | .42 |
| .470 | - .6234 | .80 | .31 |
| .693 | - .4406 | 1.00 | .25 |
| .956 | - .2768 | 1.30 | .19 |
| 1.224 | - .1573 | 1.70 | .15 |
| 1.435 | - .0932 | 2.10 | .12 |
| 1.723 | - .0392 | 2.80 | .09 |
| 2.197 | - .0056 | 4.50 | .01 |

s = 3

| u | $He_1^{(1)}(-|s|,u)$ | y | x |
|---|---|---|---|
| .000 | -.4679 | .866 | .866 |
| .038 | -.4406 | .90 | .83 |
| .144 | -.3748 | 1.00 | .75 |
| .326 | -.2770 | 1.20 | .62 |
| .480 | -.2082 | 1.40 | .54 |
| .614 | -.1590 | 1.60 | .47 |
| .732 | -.1224 | 1.80 | .42 |
| .837 | -.0950 | 2.00 | .38 |
| 1.019 | -.0582 | 2.40 | .31 |
| 1.099 | -.0457 | 2.60 | .29 |
| 1.466 | -.0121 | 3.75 | .20 |

s = 6

| u | $He_1^{(1)}(-|s|,u)$ | y | x |
|---|---|---|---|
| .000 | -.2162 | 1.225 | 1.225 |
| .060 | -.1948 | 1.30 | 1.154 |
| .203 | -.1492 | 1.50 | 1.00 |
| .490 | -.0803 | 2.00 | .75 |
| .714 | -.0446 | 2.50 | .60 |
| .827 | -.0315 | 2.80 | .54 |
| .896 | -.0252 | 3.00 | .50 |
| .991 | -.0180 | 3.30 | .45 |
| 1.106 | -.0115 | 3.70 | .40 |
| 1.407 | -.0028 | 5.00 | .30 |

s = 9

| u | $He_1^{(1)}(-|s|,u)$ | y | x |
|---|---|---|---|
| .000 | -.1227 | 1.50 | 1.50 |
| .064 | -.1086 | 1.60 | 1.41 |
| .182 | -.0853 | 1.80 | 1.25 |
| .288 | -.0675 | 2.00 | 1.12 |
| .470 | -.0427 | 2.40 | .94 |
| .550 | -.0342 | 2.60 | .87 |
| .693 | -.0220 | 3.00 | .75 |
| .903 | -.0102 | 3.70 | .61 |
| .981 | -.0074 | 4.00 | .56 |
| 1.099 | -.0043 | 4.50 | .50 |

s = 15

| u | $He_1^{(1)}(-|s|,u)$ | y | x |
|---|---|---|---|
| .000 | -.0514 | 1.936 | 1.936 |
| .032 | -.0478 | 2.00 | 1.875 |
| .255 | -.0279 | 2.50 | 1.50 |
| .438 | -.0166 | 3.00 | 1.25 |
| .725 | -.0059 | 4.00 | .94 |

$$He_2^{(1)}(-|s|,u) = i\sqrt{\frac{2}{\pi}} \left(\frac{1}{De_o}\right) \sum_{m=o}^{\infty} De_{2m} \cdot I_m(x) \cdot K_m(y) ;$$

$$y = \frac{\sqrt{|s|}}{2} e^u , \quad x = \frac{\sqrt{|s|}}{2} e^{-u}$$

| | s = 1 | | | | s = 3 | | |
|---|---|---|---|---|---|---|---|
| u | $He_2^{(1)}(-|s|,u)$ | y | x | u | $He_2^{(1)}(-|s|,u)$ | y | x |
| .000 | 6.2090i | .500 | .500 | .000 | 2.0500i | .866 | .866 |
| .182 | 4.2108i | .60 | .42 | .038 | 1.8576i | .90 | .83 |
| .470 | 2.1903i | .80 | .31 | .144 | 1.4370i | 1.00 | .75 |
| .693 | 1.3120i | 1.00 | .25 | .326 | .8987i | 1.20 | .62 |
| .956 | .6797i | 1.30 | .19 | .480 | .6045i | 1.40 | .54 |
| 1.224 | .3348i | 1.70 | .15 | .614 | .4153i | 1.60 | .47 |
| 1.435 | .1754i | 2.10 | .12 | .732 | .2953i | 1.80 | .42 |
| 1.723 | .0646i | 2.80 | .09 | .837 | .2143i | 2.00 | .38 |
| 2.197 | .0055i | 4.50 | .01 | 1.019 | .1158i | 2.40 | .31 |
| | | | | 1.099 | .0878i | 2.60 | .29 |
| | | | | 1.466 | .0194i | 3.75 | .20 |

| | s = 6 | | | | s = 9 | | |
|---|---|---|---|---|---|---|---|
| u | $He_2^{(1)}(-|s|,u)$ | y | x | u | $He_2^{(1)}(-|s|,u)$ | y | x |
| .000 | 1.0535i | 1.225 | 1.225 | .000 | .7248i | 1.500 | 1.500 |
| .060 | .8856i | 1.30 | 1.15 | .064 | .5903i | 1.60 | 1.41 |
| .203 | .5856i | 1.50 | 1.00 | .182 | .3997i | 1.80 | 1.25 |
| .490 | .2398i | 2.00 | .75 | .288 | .2786i | 2.00 | 1.12 |
| .714 | .1110i | 2.50 | .60 | .470 | .1463i | 2.40 | .94 |
| .827 | .0727i | 2.80 | .54 | .550 | .1083i | 2.60 | .87 |
| .896 | .0550i | 3.00 | .50 | .693 | .0608i | 3.00 | .75 |
| .991 | .0367i | 3.30 | .45 | .903 | .0240i | 3.70 | .61 |
| 1.106 | .0219i | 3.70 | .40 | .981 | .0163i | 4.00 | .56 |
| 1.407 | .0045i | 5.00 | .30 | 1.099 | .0088i | 4.50 | .50 |

| | s = 15 | | | | s = 25 | | |
|---|---|---|---|---|---|---|---|
| u | $He_2^{(1)}(-|s|,u)$ | y | x | u | $He_2^{(1)}(-|s|,u)$ | y | x |
| .000 | .4394i | 1.936 | 1.936 | .000 | .2294i | 2.500 | 2.500 |
| .032 | .3969i | 2.00 | 1.88 | .182 | .0978i | 3.00 | 2.08 |
| .255 | .1616i | 2.50 | 1.50 | .470 | .0228i | 4.00 | 1.56 |
| .438 | .0748i | 3.00 | 1.25 | .693 | .0063i | 5.00 | 1.25 |
| .725 | .0190i | 4.00 | .94 | | | | |

$$Ho_1^{(1)}(-|s|,u) = -\sqrt{\frac{2}{\pi}} \left(\frac{1}{De_1}\right) \sum_{m=o}^{\infty} De_{2m+1}\left[I_{m+1}(x)\cdot K_m(y) + I_m(x)\cdot K_{m+1}(y)\right] ;$$

$$y = \frac{\sqrt{|s|}}{2} e^u, \quad x = \frac{\sqrt{|s|}}{2} e^{-u}$$

s = 1

| u | $Ho_1^{(1)}(-|s|,u)$ | y | x |
|---|---|---|---|
| .000 | −1.54487 | .500 | .500 |
| .182 | −1.19 | .60 | .42 |
| .470 | − .76 | .80 | .31 |
| .693 | − .52 | 1.00 | .25 |
| .956 | − .32 | 1.30 | .19 |
| 1.224 | − .18 | 1.70 | .15 |
| 1.435 | − .10 | 2.10 | .12 |
| 1.723 | − .04 | 2.80 | .09 |
| 2.197 | − .01 | 4.50 | .01 |

s = 3

| u | $Ho_1^{(1)}(-|s|,u)$ | y | x |
|---|---|---|---|
| .000 | −.82970 | .866 | .866 |
| .144 | −.63 | 1.00 | .75 |
| .326 | −.43 | 1.20 | .62 |
| .480 | −.31 | 1.40 | .54 |
| .614 | −.23 | 1.60 | .47 |
| .732 | −.17 | 1.80 | .42 |
| .837 | −.13 | 2.00 | .38 |
| 1.019 | −.08 | 2.40 | .31 |
| 1.099 | −.06 | 2.60 | .29 |
| 1.466 | −.01 | 3.75 | .20 |

s = 6

| u | $Ho_1^{(1)}(-|s|,u)$ | y | x |
|---|---|---|---|
| .000 | −.51563 | 1.225 | 1.225 |
| .060 | −.45 | 1.30 | 1.154 |
| .203 | −.31 | 1.50 | 1.00 |
| .490 | −.14 | 2.00 | .75 |
| .714 | −.07 | 2.50 | .60 |
| .827 | −.05 | 2.80 | .54 |
| .896 | −.04 | 3.00 | .50 |
| .991 | −.03 | 3.30 | .45 |
| 1.106 | −.02 | 3.70 | .40 |
| 1.407 | −.00 | 5.00 | .30 |

s = 9

| u | $Ho_1^{(1)}(-|s|,u)$ | y | x |
|---|---|---|---|
| .000 | −.36078 | 1.500 | 1.500 |
| .064 | −.30 | 1.60 | 1.41 |
| .182 | −.22 | 1.80 | 1.25 |
| .288 | −.16 | 2.00 | 1.12 |
| .470 | −.09 | 2.40 | .94 |
| .550 | −.07 | 2.60 | .87 |
| .693 | −.04 | 3.00 | .75 |
| .903 | −.02 | 3.70 | .61 |
| .981 | −.01 | 4.00 | .56 |
| 1.099 | −.01 | 4.50 | .50 |

s = 15

| u | $Ho_1^{(1)}(-|s|,u)$ | y | x |
|---|---|---|---|
| .000 | −.19351 | 1.936 | 1.936 |
| .032 | −.17 | 2.00 | 1.875 |
| .255 | −.08 | 2.50 | 1.50 |
| .438 | −.04 | 3.00 | 1.25 |
| .725 | −.01 | 4.00 | .94 |

$$Ho_2^{(1)}(-|s|,u) = i\sqrt{\frac{2}{\pi}}\left(\frac{1}{Do_2}\right)\sum_{m=o}^{\infty} Do_{2m}\left[I_{m-1}(x)\cdot K_{m+1}(y) - I_{m+1}(x)\cdot K_{m-1}(y)\right];$$

$$y = \frac{\sqrt{|s|}}{2}e^u, \quad x = \frac{\sqrt{|s|}}{2}e^{-u}$$

s = 1

| u | $Ho_2^{(1)}(-|s|,u)$ | y | x |
|---|---|---|---|
| .000 | 6.12028i | .500 | .500 |
| .182 | 4.128i | .60 | .42 |
| .470 | 2.180i | .80 | .31 |
| .693 | 1.299i | 1.00 | .25 |
| .956 | .680i | 1.30 | .19 |
| 1.224 | .328i | 1.70 | .15 |
| 1.435 | .174i | 2.10 | .12 |
| 1.723 | .064i | 2.80 | .09 |
| 2.197 | .008i | 4.50 | .01 |

s = 3

| u | $Ho_2^{(1)}(-|s|,u)$ | y | x |
|---|---|---|---|
| .000 | 1.87154i | .866 | .866 |
| .038 | 1.705i | .90 | .83 |
| .144 | 1.323i | 1.00 | .75 |
| .326 | .839i | 1.20 | .62 |
| .480 | .563i | 1.40 | .54 |
| .614 | .390i | 1.60 | .47 |
| .732 | .278i | 1.80 | .42 |
| .837 | .202i | 2.00 | .38 |
| 1.019 | .111i | 2.40 | .31 |
| 1.099 | .084i | 2.60 | .29 |
| 1.466 | .019i | 3.75 | .20 |

s = 6

| u | $Ho_2^{(1)}(-|s|,u)$ | y | x |
|---|---|---|---|
| .000 | .81826i | 1.225 | 1.225 |
| .060 | .697i | 1.30 | 1.154 |
| .203 | .470i | 1.50 | 1.00 |
| .490 | .200i | 2.00 | .75 |
| .714 | .095i | 2.50 | .60 |
| .827 | .063i | 2.80 | .54 |
| .896 | .048i | 3.00 | .50 |
| .991 | .033i | 3.30 | .45 |
| 1.106 | .020i | 3.70 | .40 |
| 1.407 | .004i | 5.00 | .30 |

s = 9

| u | $Ho_2^{(1)}(-|s|,u)$ | y | x |
|---|---|---|---|
| .000 | .47471i | 1.500 | 1.500 |
| .064 | .393i | 1.60 | 1.41 |
| .182 | .275i | 1.80 | 1.25 |
| .288 | .198i | 2.00 | 1.12 |
| .470 | .108i | 2.40 | .94 |
| .550 | .081i | 2.60 | .87 |
| .693 | .047i | 3.00 | .75 |
| .903 | .019i | 3.70 | .61 |
| .981 | .013i | 4.00 | .56 |
| 1.099 | .007i | 4.50 | .50 |

s = 15

| u | $Ho_2^{(1)}(-|s|,u)$ | y | x |
|---|---|---|---|
| .000 | .21356i | 1.936 | 1.936 |
| .032 | .192i | 2.00 | 1.875 |
| .255 | .089i | 2.50 | 1.50 |
| .438 | .044i | 3.00 | 1.25 |
| .725 | .013i | 4.00 | .94 |

s = 25

| u | $Ho_2^{(1)}(-|s|,u)$ | y | x |
|---|---|---|---|
| .000 | .07841i | 2.50 | 2.50 |
| .182 | .038i | 3.00 | 2.08 |
| .470 | .011i | 4.00 | 1.56 |
| .693 | .003i | 5.00 | 1.25 |

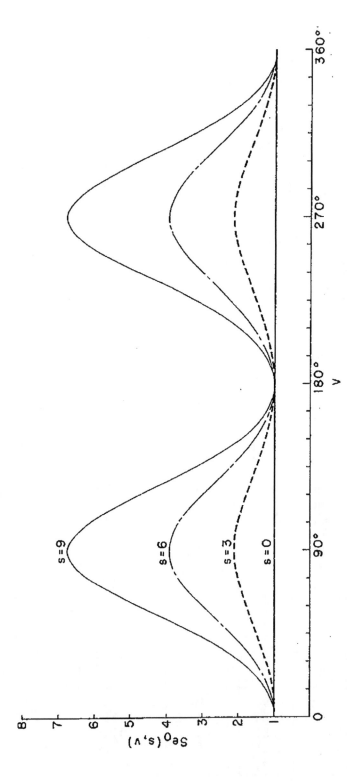

FIGURE 2  EVEN PERIODIC MATHIEU FUNCTION OF ORDER ZERO

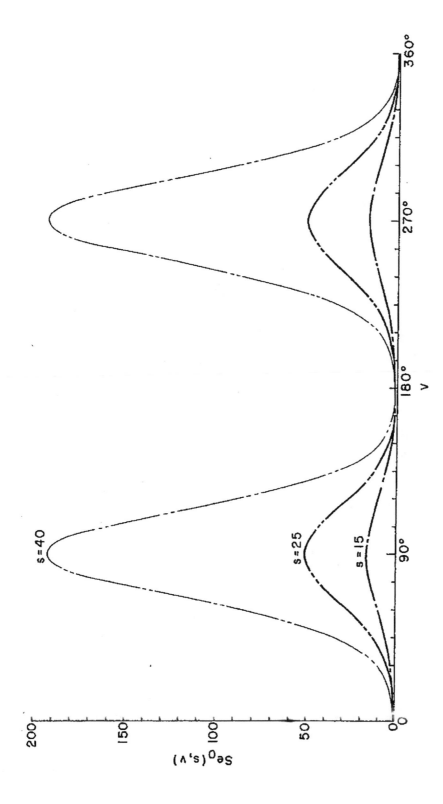

FIGURE 3  EVEN PERIODIC MATHIEU FUNCTION OF ORDER ZERO

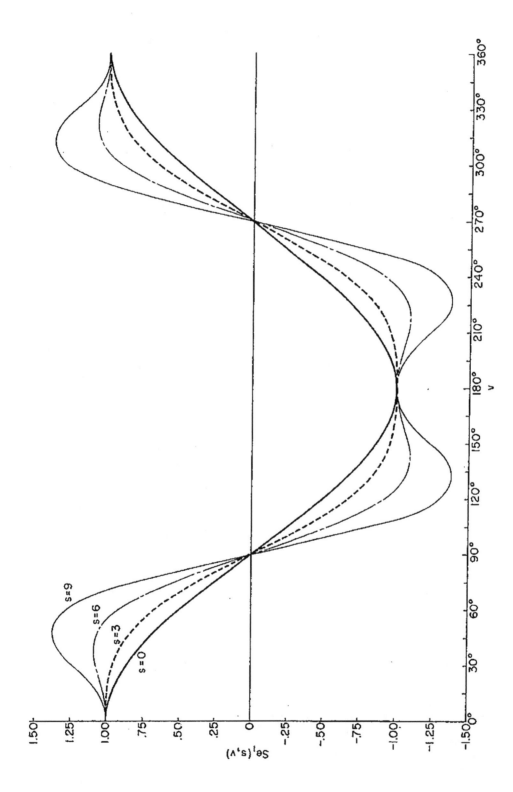

FIGURE 4   EVEN PERIODIC MATHIEU FUNCTION OF ORDER ONE

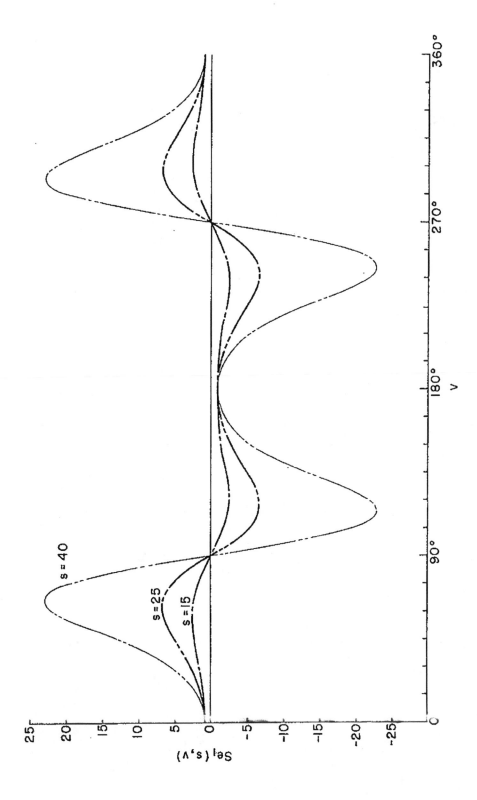

FIGURE 5 EVEN PERIODIC MATHIEU FUNCTION OF ORDER ONE

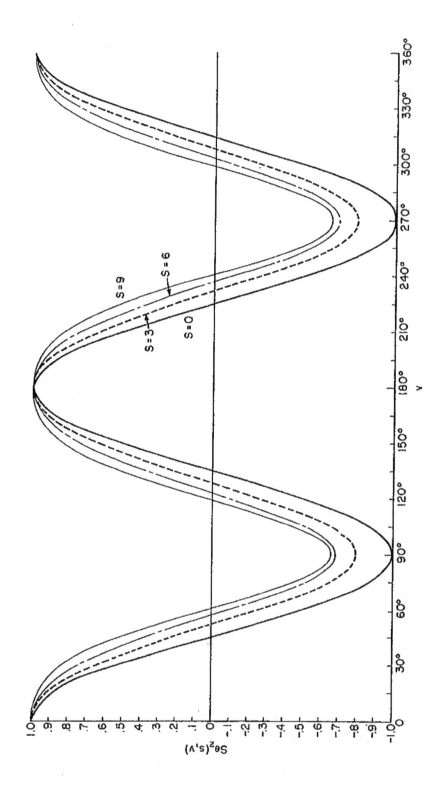

FIGURE 6  EVEN PERIODIC MATHIEU FUNCTION OF ORDER TWO

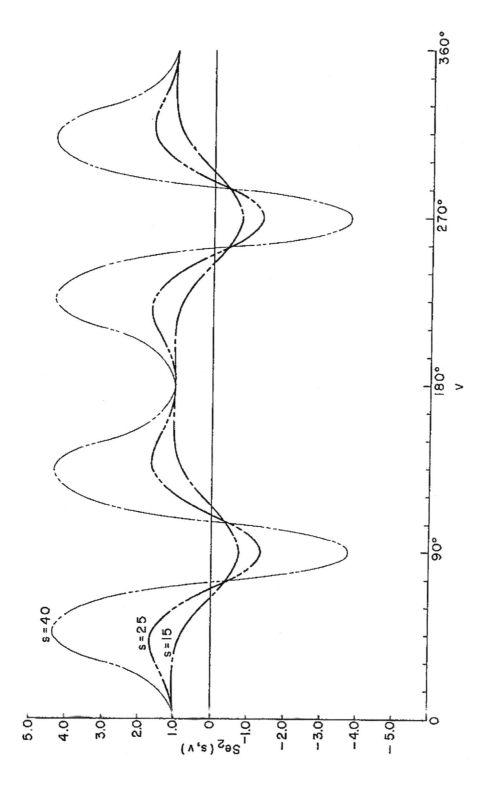

FIGURE 7 EVEN PERIODIC MATHIEU FUNCTION OF ORDER TWO

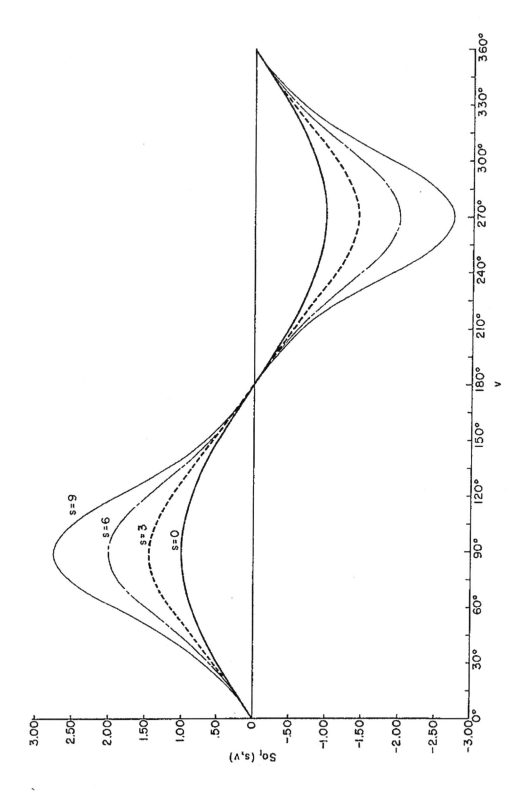

FIGURE 8  ODD PERIODIC MATHIEU FUNCTION OF ORDER ONE

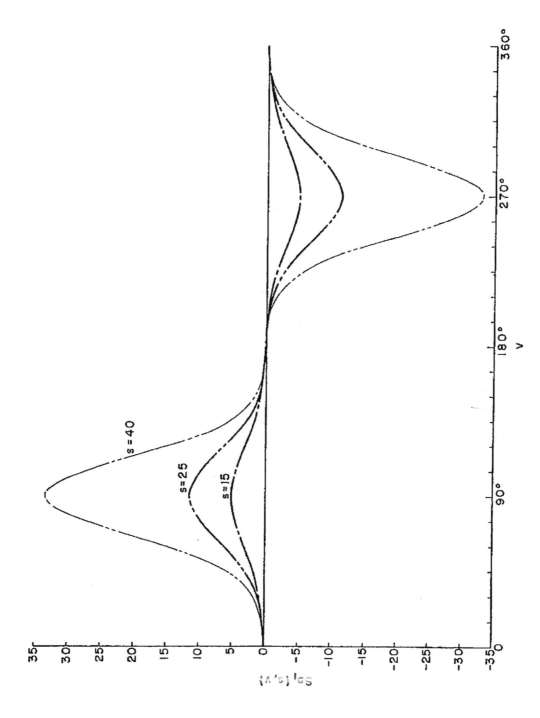

FIGURE 9  ODD PERIODIC MATHIEU FUNCTION OF ORDER ONE

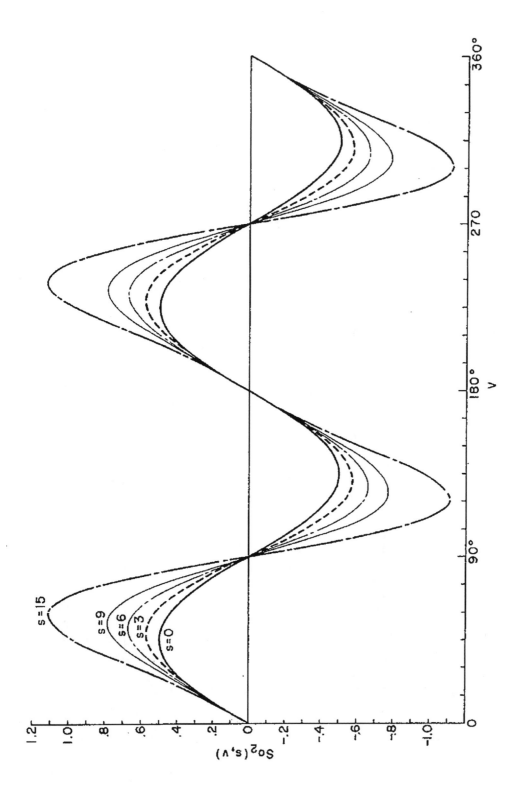

FIGURE 10  ODD PERIODIC MATHIEU FUNCTION OF ORDER TWO

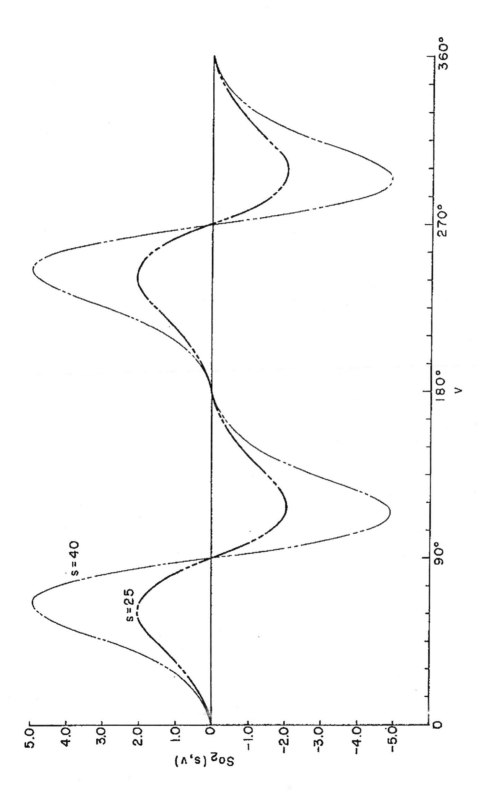

FIGURE 11  ODD PERIODIC MATHIEU FUNCTION OF ORDER TWO

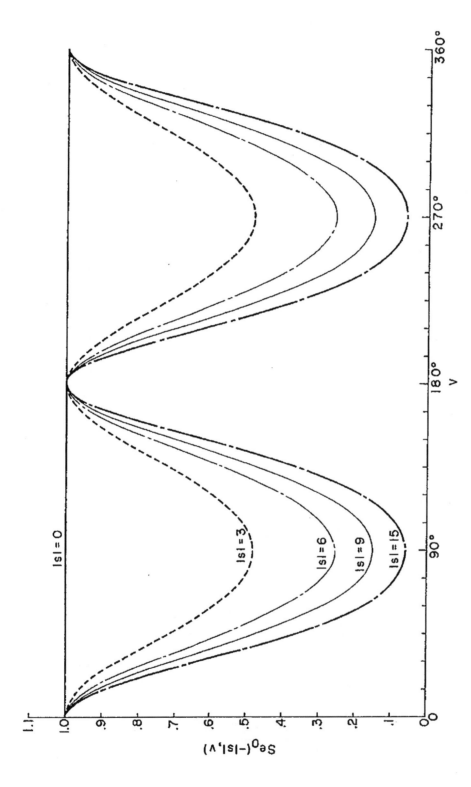

FIGURE 12   EVEN PERIODIC MATHIEU FUNCTION OF ORDER ZERO AND
NEGATIVE PARAMETER "s"

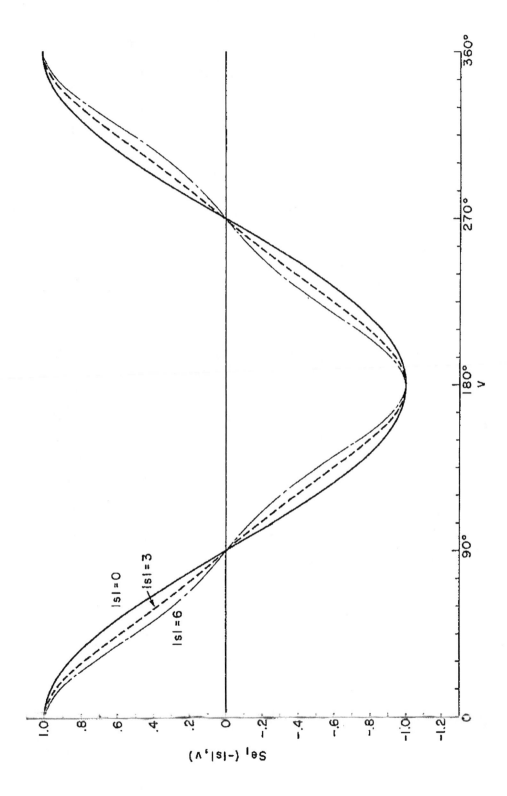

FIGURE 13   EVEN PERIODIC MATHIEU FUNCTION OF ORDER ONE AND
NEGATIVE PARAMETER "s"

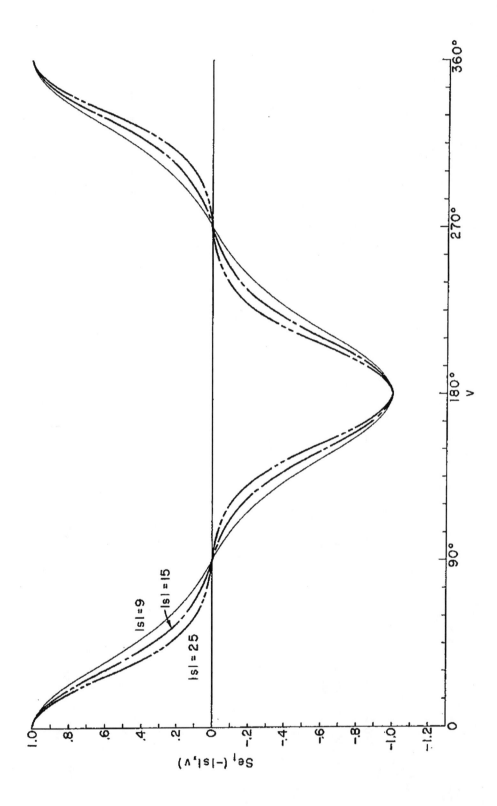

FIGURE 14  EVEN PERIODIC MATHIEU FUNCTION OF ORDER ONE AND
NEGATIVE PARAMETER "s"

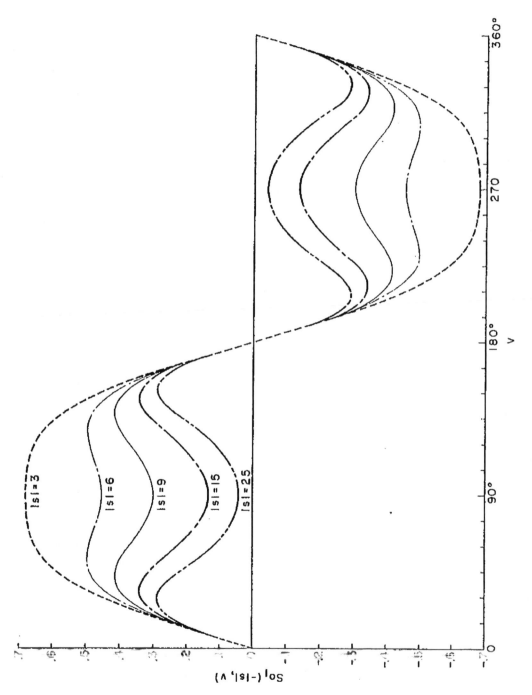

FIGURE 15   ODD PERIODIC MATHIEU FUNCTION OF ORDER ONE AND
NEGATIVE PARAMETER "s"

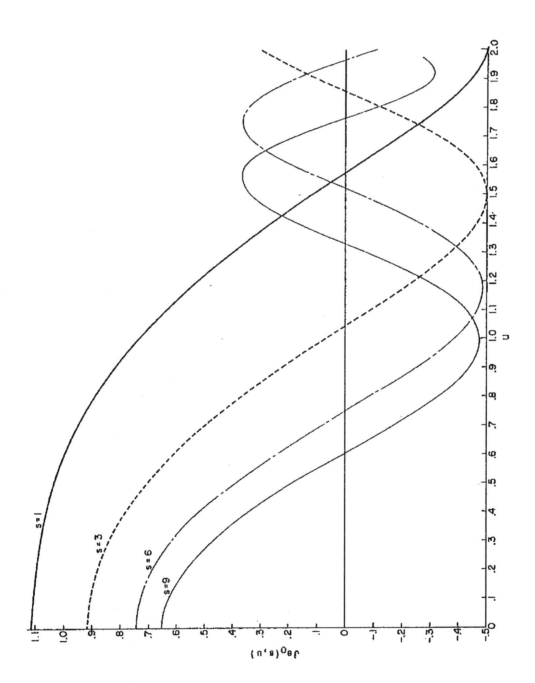

FIGURE 16  "EVEN" RADIAL MATHIEU FUNCTION OF THE FIRST KIND
AND ORDER ZERO

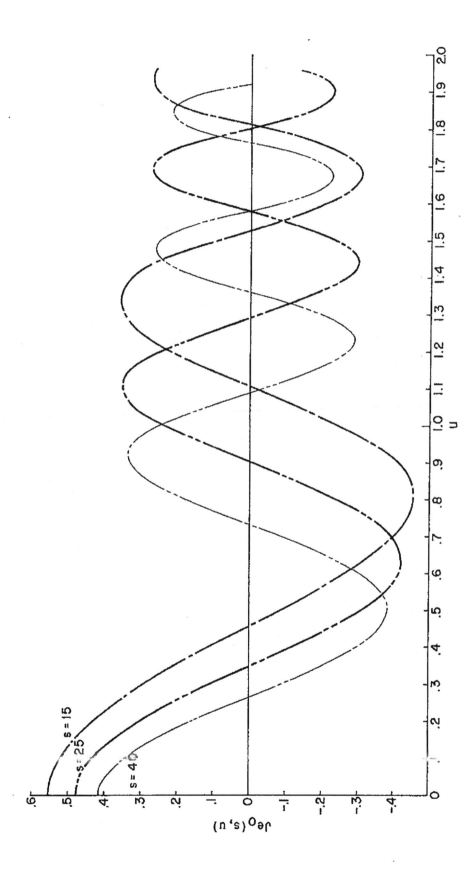

FIGURE 17  "EVEN" RADIAL MATHIEU FUNCTION OF THE FIRST KIND
AND ORDER ZERO

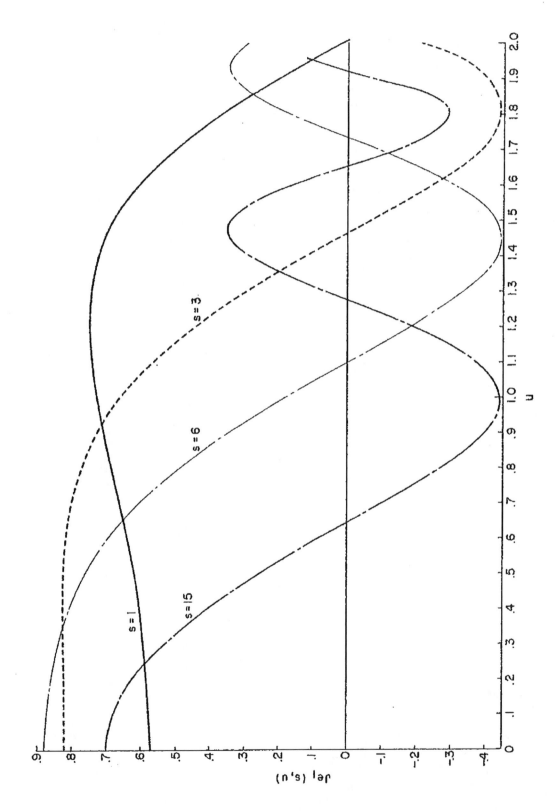

FIGURE 18 "EVEN" RADIAL MATHIEU FUNCTION OF THE FIRST KIND
AND ORDER ONE

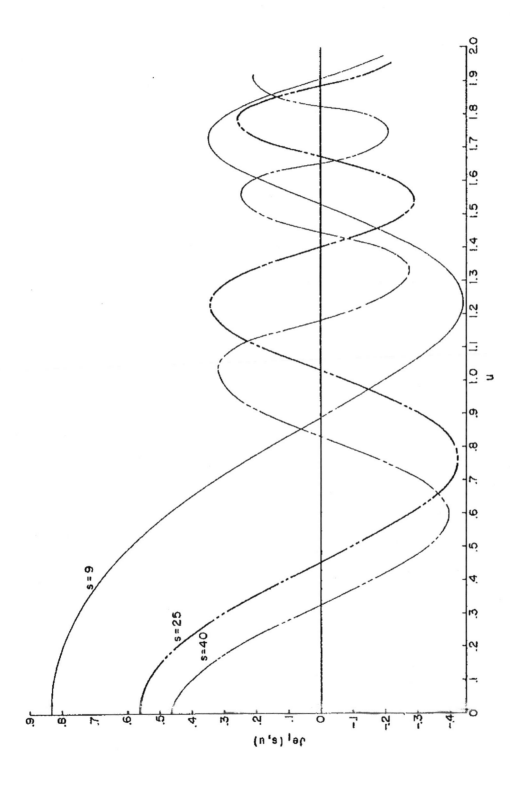

FIGURE 19 "EVEN" RADIAL MATHIEU FUNCTION OF THE FIRST KIND AND ORDER ONE

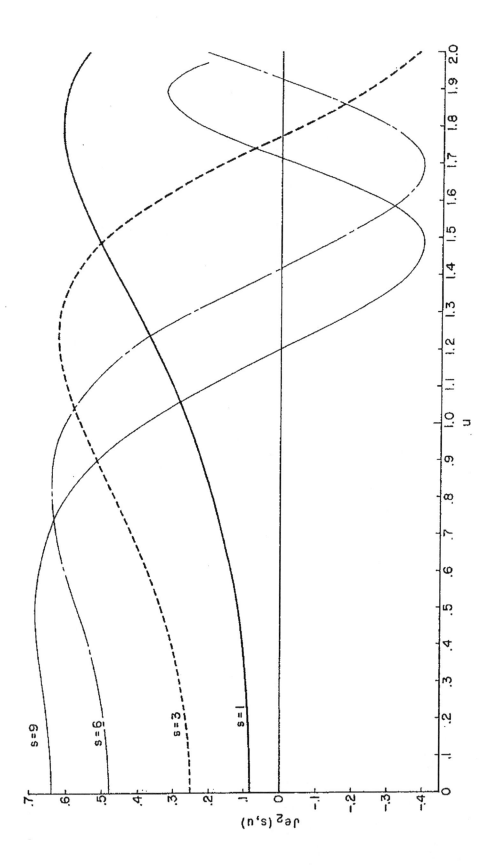

FIGURE 20 "EVEN" RADIAL MATHIEU FUNCTION OF THE FIRST KIND
AND ORDER TWO

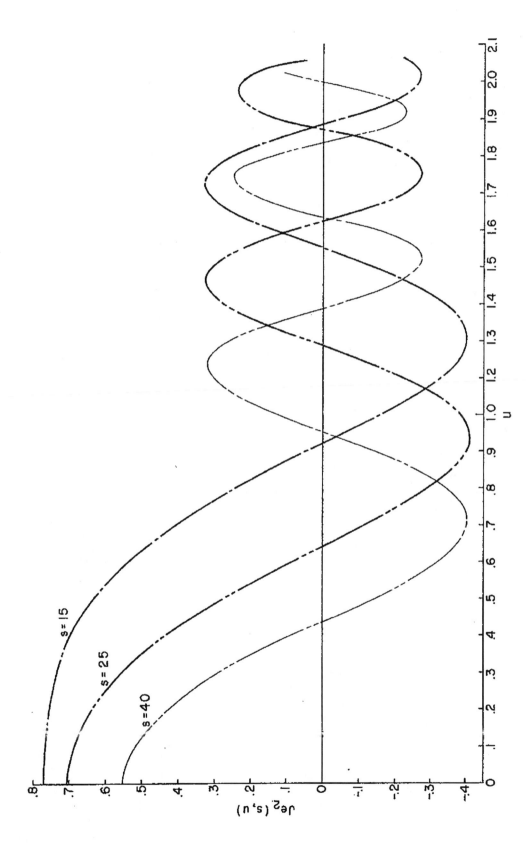

FIGURE 21   "EVEN" RADIAL MATHIEU FUNCTION OF THE FIRST KIND
AND ORDER TWO

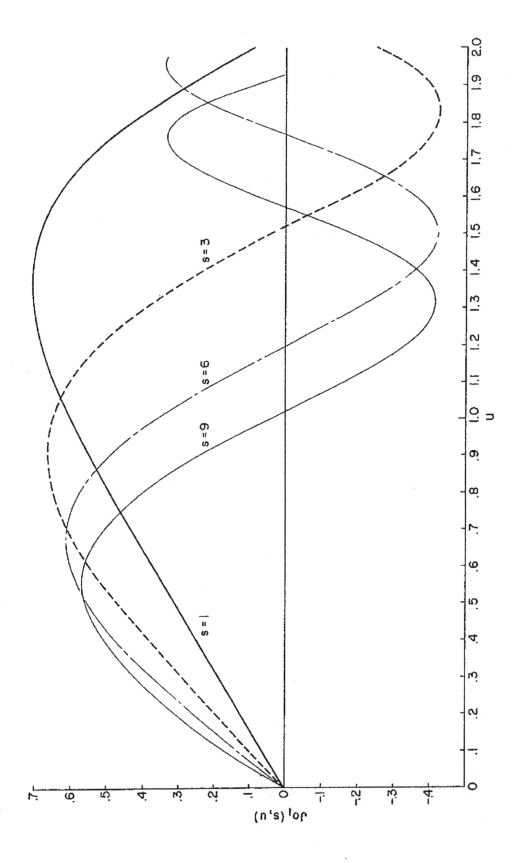

FIGURE 22  "ODD" RADIAL MATHIEU FUNCTION OF THE FIRST KIND
AND ORDER ONE

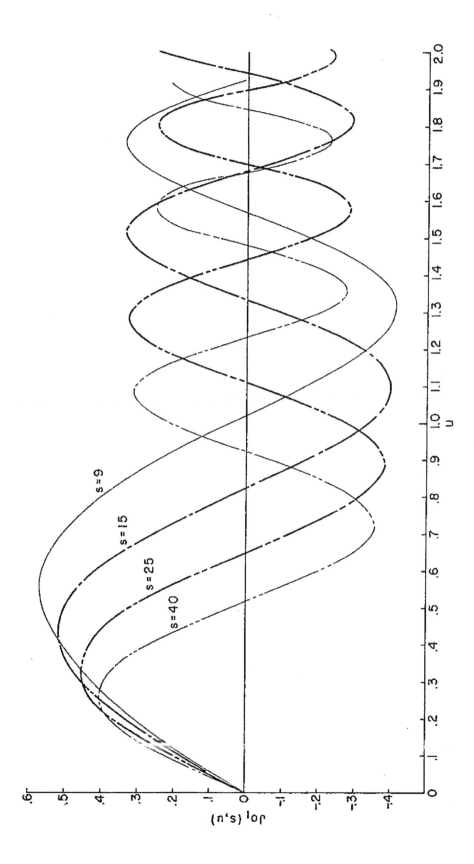

FIGURE 23 "ODD" RADIAL MATHIEU FUNCTION OF THE FIRST KIND
AND ORDER ONE

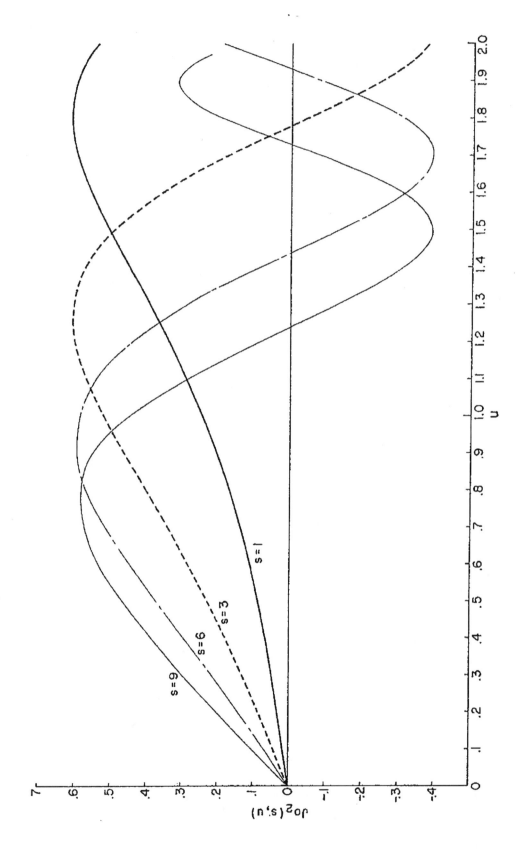

FIGURE 24 "ODD" RADIAL MATHIEU FUNCTION OF THE FIRST KIND
AND ORDER TWO

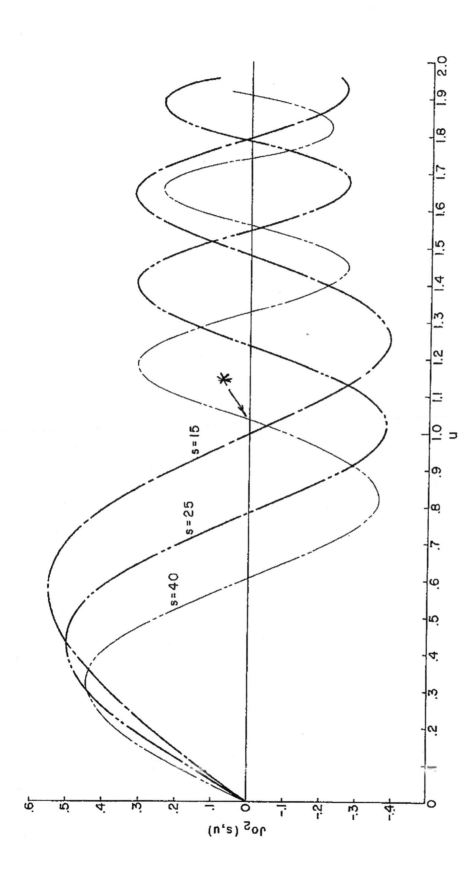

FIGURE 25 "ODD" RADIAL MATHIEU FUNCTION OF THE FIRST KIND
AND ORDER TWO

$$Jo_2(s=40, u) = 0 \text{ at } u = 1.025$$

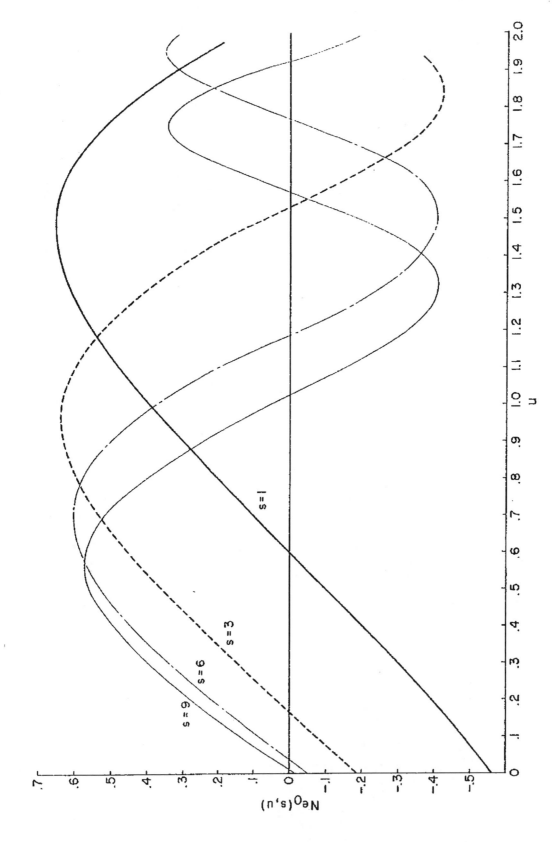

FIGURE 26  "EVEN" RADIAL MATHIEU FUNCTION OF THE SECOND KIND
AND ORDER ZERO

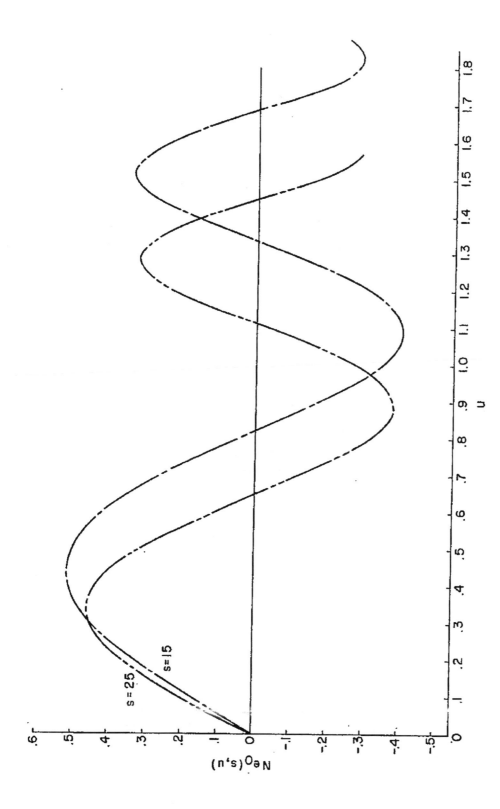

FIGURE 27  "EVEN" RADIAL MATHIEU FUNCTION OF THE SECOND KIND
AND ORDER ZERO

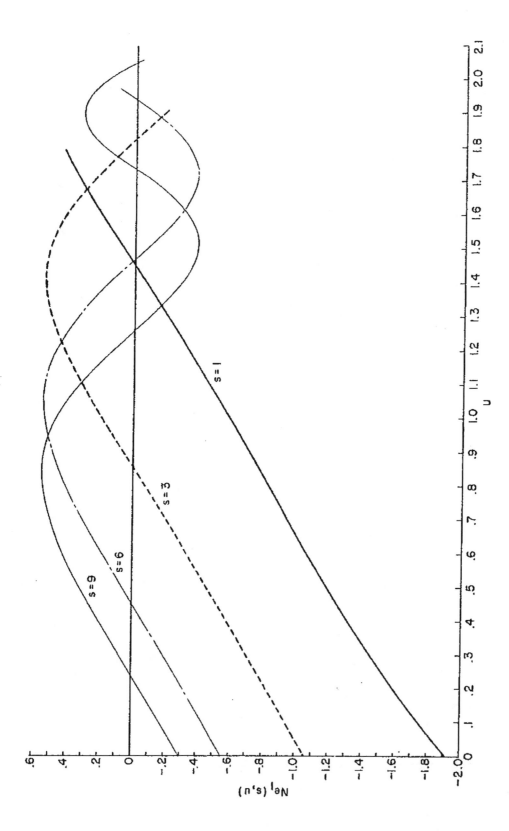

FIGURE 28   "EVEN" RADIAL MATHIEU FUNCTION OF THE SECOND KIND
AND ORDER ONE

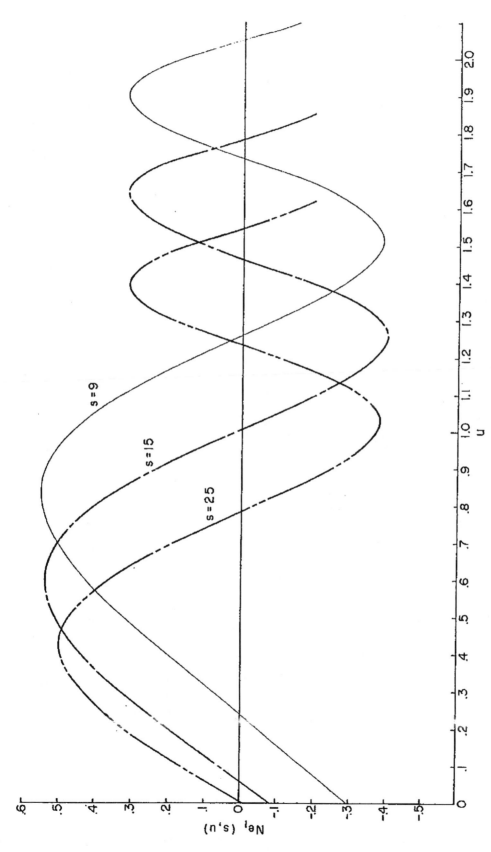

FIGURE 29  "EVEN" RADIAL MATHIEU FUNCTION OF THE SECOND KIND
AND ORDER ONE

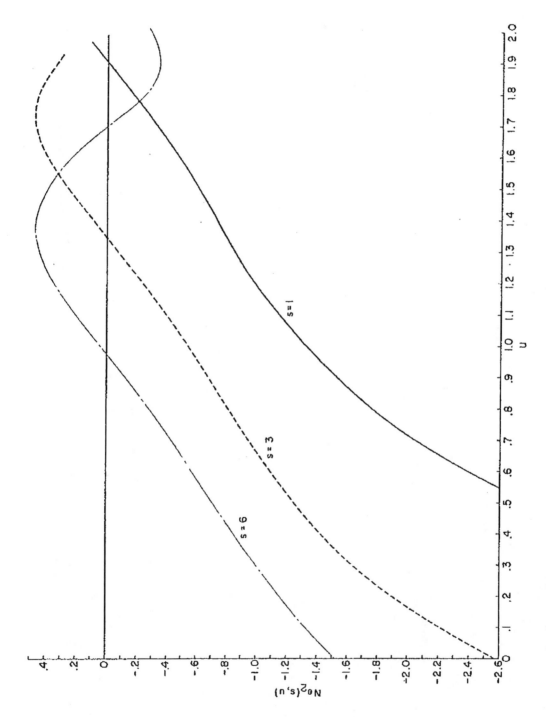

FIGURE 30   "EVEN" RADIAL MATHIEU FUNCTION OF THE SECOND KIND
AND ORDER TWO

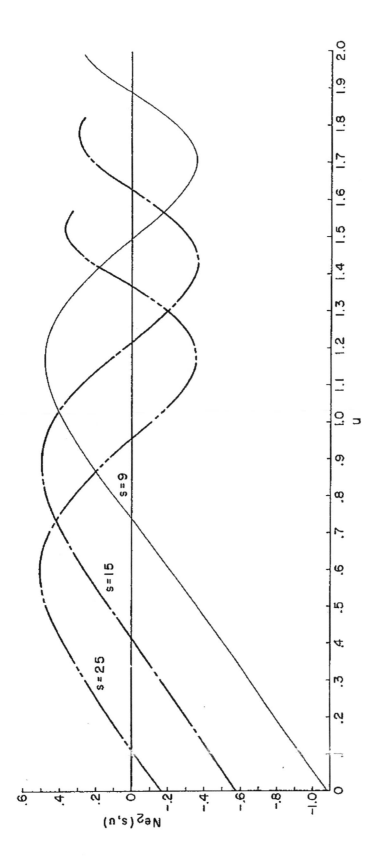

FIGURE 31   "EVEN" RADIAL MATHIEU FUNCTION OF THE SECOND KIND
AND ORDER TWO

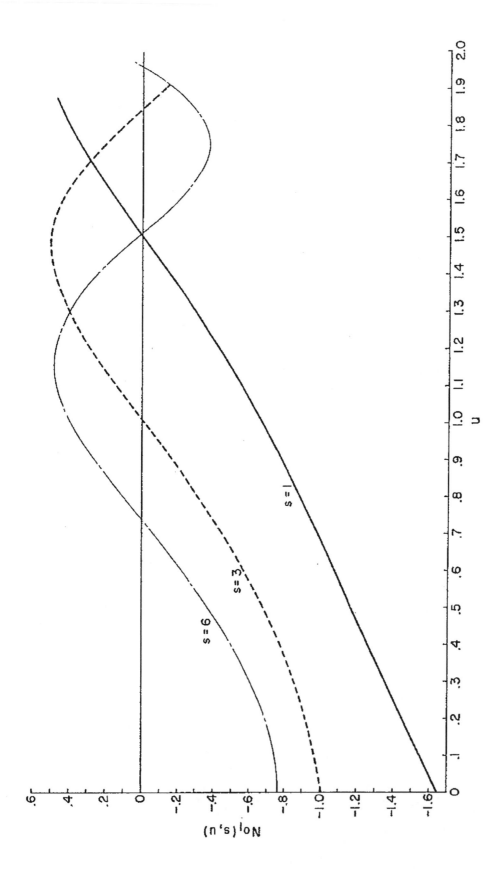

FIGURE 32 "ODD" RADIAL MATHIEU FUNCTION OF THE SECOND KIND
AND ORDER ONE

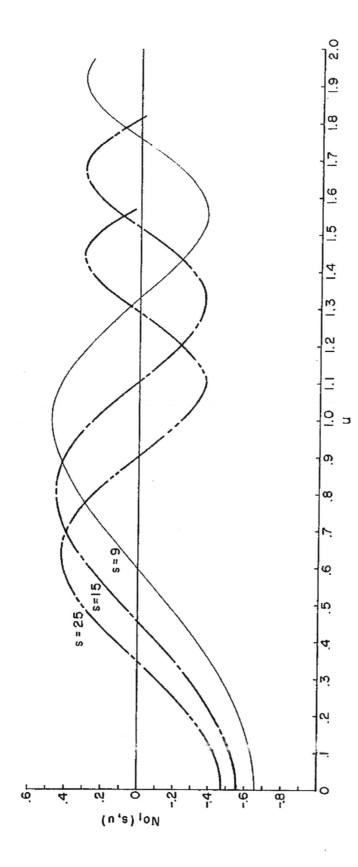

FIGURE 33 "ODD" RADIAL MATHIEU FUNCTION OF THE SECOND KIND
AND ORDER ONE

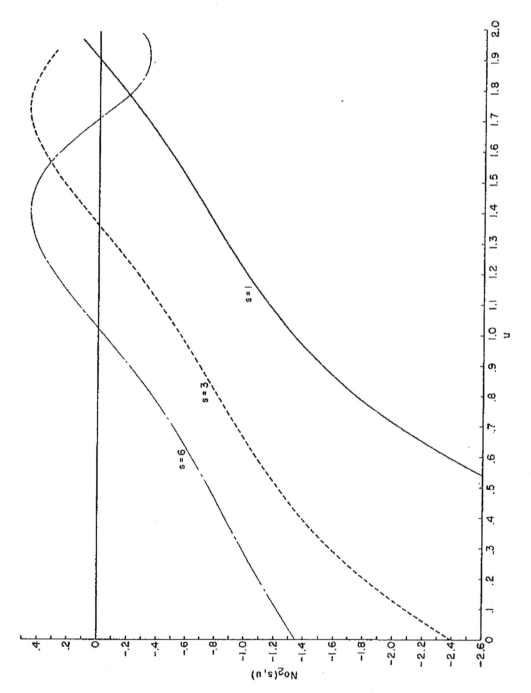

FIGURE 34 "ODD" RADIAL MATHIEU FUNCTION OF THE SECOND KIND
AND ORDER TWO

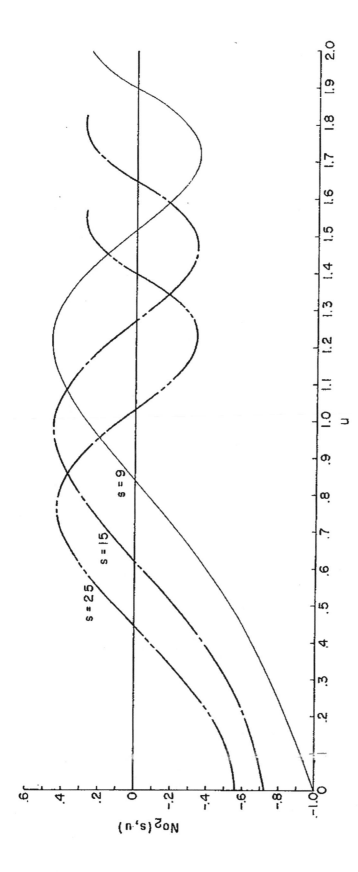

FIGURE 35    "ODD" RADIAL MATHIEU FUNCTION OF THE SECOND KIND
AND ORDER TWO

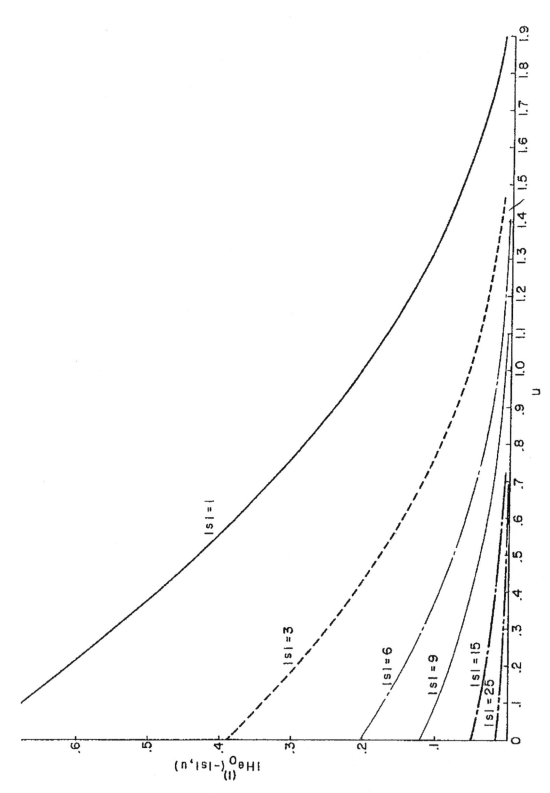

FIGURE 36   "EVEN" RADIAL MATHIEU FUNCTION OF THE THIRD KIND AND ORDER
ZERO, FOR NEGATIVE REAL VALUES OF THE PARAMETER "s"

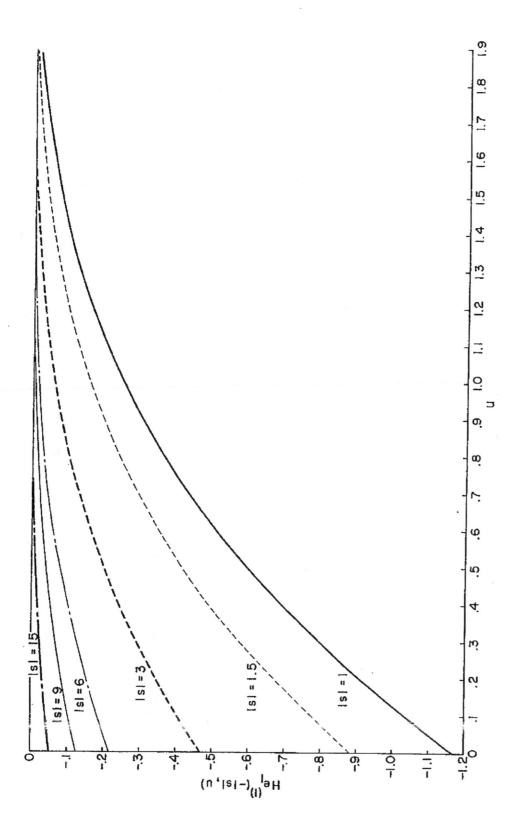

FIGURE 37 "EVEN" RADIAL MATHIEU FUNCTION OF THE THIRD KIND AND ORDER ONE, FOR NEGATIVE REAL VALUES OF THE PARAMETER "s"

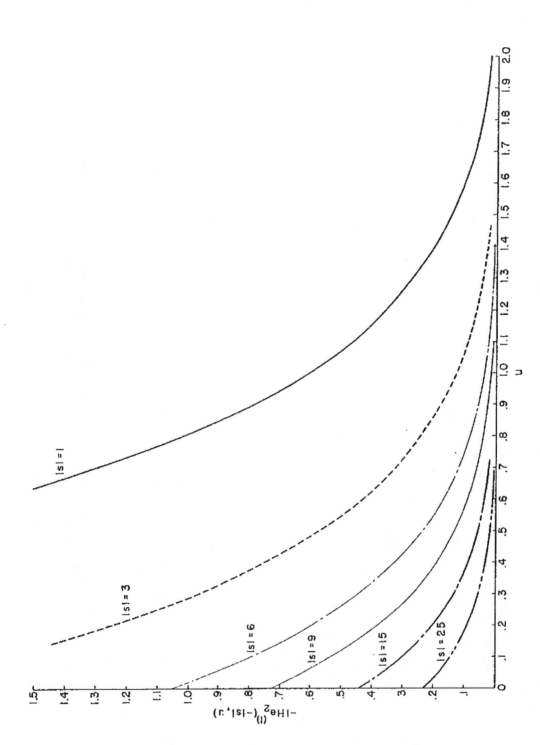

FIGURE 38   "EVEN" RADIAL MATHIEU FUNCTION OF THE THIRD KIND AND ORDER
TWO, FOR NEGATIVE REAL VALUES OF THE PARAMETER "s"

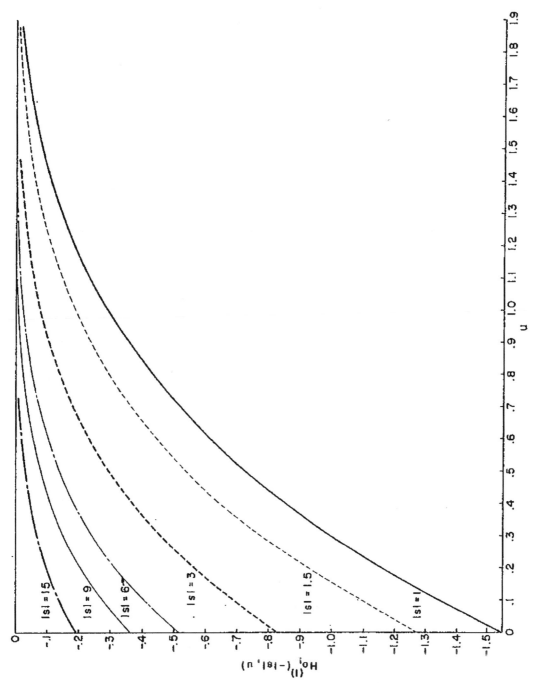

FIGURE 39   "ODD" RADIAL MATHIEU FUNCTION OF THE THIRD KIND AND, ORDER
ONE, FOR NEGATIVE REAL VALUES OF THE PARAMETER "s"

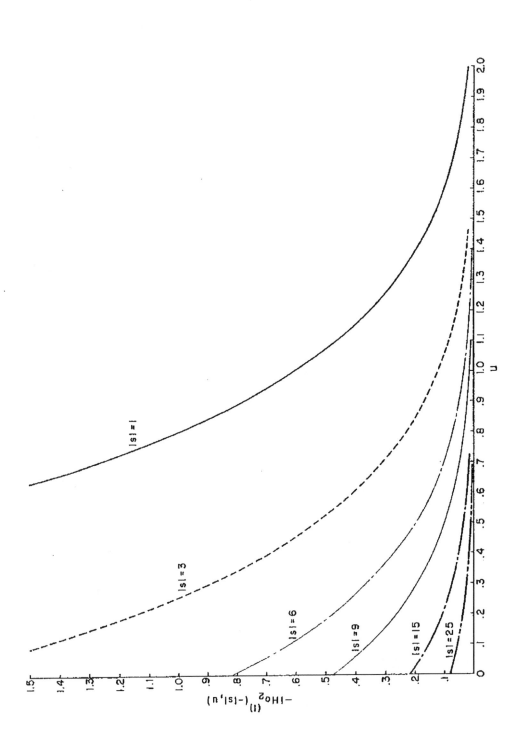

FIGURE 40  "ODD" RADIAL MATHIEU FUNCTION OF THE THIRD KIND AND ORDER TWO, FOR NEGATIVE REAL VALUES OF THE PARAMETER "s"

# SECTION II

## Derivatives and Zeroes of Periodic and Radial Mathieu Functions

iii

## LIST OF ILLUSTRATIONS

# SUMMARY

The major portion of this section consists of tabulated values and graphs of the first derivatives of the periodic Mathieu functions $[Se_n' \ (s, v) \ and \ So_n' \ (s, v)]$ and the radial Mathieu functions of the first, second, and third kinds

$[Je_n' \ (s, u), Jo_n' \ (s, u), Ne_n' \ (s, u), No_n' \ (s, u), He_n^{(1)'} \ (s, u), and \ Ho_n^{(1)'}(s, u)]$ for orders zero, one, and two and for various positive and/or negative values of the parameters. Tables of the first and second zeros of the radial functions, $Je_n \ (s, u)$, and the first zeros of the first derivatives of these functions are also included. In addition, the curves and data for the periodic functions $Se_2$ (s, v) and $So_2 \ (s, v)$ are given for various negative real values of the parameters.

# INTRODUCTION

Many physical problems which utilize Mathieu functions also require their first derivatives. Values of the Mathieu functions of orders zero, one, and two have been given previously[1], and the present report contains data and curves for the first derivatives of these same functions. The positions of zeros for the radial functions of the first kind ($Je_n$ and $Jo_n$) and for the first derivatives of these functions also have been calculated and are tabulated herein. (Among other uses, the zeros are necessary for determining cut off frequencies of modes in hollow metal waveguides of elliptical cross-section[2,3].) In addition, curves and data are included for the even and odd periodic Mathieu functions of order two and negative real parameter s $\left[\text{i.e.,} \right.$ $Se_2(-|s|, v)$ and $So_2(-|s|, v)$ as defined in Reference $4\Big]$.

## General Background

The Mathieu equation may be written as follows:

$$\frac{\partial^2 g}{\partial v^2} + (b - s \cos^2 v) g = 0 , \tag{1}$$

and the modified Mathieu equation is, similarly:

$$\frac{\partial^2 f}{\partial u^2} - (b - s \cosh^2 u) f = 0 , \tag{2}$$

where b is the separation constant or characteristic value. Equation (2) may be obtained from Equation (1) by replacing v by iu . Only real values of s and b are considered in these reports.

The angular solutions of period $\pi$ or $2\pi$ obtained from Equation (1) are called the periodic Mathieu functions and are designated by $Se_n$ or $So_n$, depending on whether the solutions are even or odd, respectively, in v. The periodic functions are represented as follows:

$$Se_{2r}(s, v) = \sum_{k=0}^{\infty} De_{2k}^{(2r)} \cos 2kv \qquad \text{(period } \pi) \qquad (3)$$

$$Se_{2r+1}(s, v) = \sum_{k=0}^{\infty} De_{2k+1}^{(2r+1)} \cos (2k+1) v \qquad \text{(period } 2\pi) \qquad (4)$$

$$So_{2r}(s, v) = \sum_{k=1}^{\infty} Do_{2k}^{(2r)} \sin 2kv \qquad \text{(period } \pi) \qquad (5)$$

$$So_{2r+1}(s, v) = \sum_{k=0}^{\infty} Do_{2k+1}^{(2r+1)} \sin (2k+1) v \qquad \text{(period } 2\pi) \qquad (6)$$

$De_m^{(n)}$ and $Do_m^{(n)}$ are dependent upon s and n, the order of the appropriate Mathieu function (see Reference 4, which contains a more complete discussion and tables of these coefficients). The superscripts will be omitted for simplicity, since the order is usually obvious.

The solutions of Equation (2), the modified Mathieu equation, are known as the radial Mathieu functions. The radial functions of the first kind may be expressed as follows:

$$Je_{2r+p}(s, u) = (-1)^r \sqrt{\frac{\pi}{2}} \sum_{k=0}^{\infty} (-1)^k De_{2k+p} J_{2k+p}( \sqrt{s} \cosh u) \qquad (7)$$

$$Jo_{2r+p}(s, u) = (-1)^r \sqrt{\frac{\pi}{2}} \tanh u \sum_{k=0}^{\infty} (-1)^k (2k+p) Do_{2k+p} J_{2k+p}(\sqrt{s} \cosh u),$$
$$(8)$$

where $J_{2k+p}( \sqrt{s} \cosh u)$ is the Bessel function of the first kind and of order 2k+p, p = 0 or 1.

Second independent solutions of Equation (2), or radial
Mathieu functions of the second kind, may be defined in terms of
series of Bessel function products:

$$Ne_{2r}(s, u) = (-1)^r \sqrt{\frac{\pi}{2}} \sum_{k=0}^{\infty} (-1)^k \frac{De_{2k}}{De_0} \left[ Y_k(y) \, J_k(x) \right] \tag{9}$$

$$Ne_{2r+1}(s, u) = (-1)^r \sqrt{\frac{\pi}{2}} \sum_{k=0}^{\infty} (-1)^k \frac{De_{2k+1}}{De_1} \left[ Y_{k+1}(y) \, J_k(x) + Y_k(y) \, J_{k+1}(x) \right] \tag{10}$$

$$No_{2r}(s, u) = (-1)^r \sqrt{\frac{\pi}{2}} \sum_{k=1}^{\infty} (-1)^k \frac{Do_{2k}}{Do_2} \left[ Y_{k+1}(y) \, J_{k-1}(x) - Y_{k-1}(y) \, J_{k+1}(x) \right] \tag{11}$$

$$No_{2r+1}(s, u) = (-1)^r \sqrt{\frac{\pi}{2}} \sum_{k=0}^{\infty} (-1)^k \frac{Do_{2k+1}}{Do_1} \left[ Y_{k+1}(y) \, J_k(x) - Y_k(y) \, J_{k+1}(x) \right] \tag{12}$$

where $y = \frac{\sqrt{s}}{2} e^u$, $x = \frac{\sqrt{s}}{2} e^{-u}$, and $J_m(x)$ and $Y_m(y)$ are the Bessel
functions of the first and second kinds, respectively. Except when
$s \to 0$, these series converge rapidly for real values of $u \geq 0$.

Solutions of Equation (2) analogous to the Hankel functions
have also been defined, as follows:

$$He_n^{(1)}(s, u) = Je_n + i \, Ne_n, \qquad Ho_n^{(1)}(s, u) = Jo_n + i \, No_n, \tag{13}$$

$$He_n^{(2)}(s, u) = Je_n - i \, Ne_n, \qquad Ho_n^{(2)}(s, u) = Jo_n - i \, No_n, \tag{14}$$

where Equations (13) and (14) give the radial Mathieu functions of the
third and fourth kinds, respectively. When the parameter $s$ is a
negative real quantity, the radial solution which has the proper behavior at infinity for many practical applications is the Hankel-like
Mathieu function of the third kind.

The following expressions were used for computation:

$$He_{2r}^{(1)}(-|s|, u) = i(-1)^{r+1} \sqrt{\frac{2}{\pi}} \sum_{m=0}^{\infty} \frac{De_{2m}}{De_0} I_m(x) K_m(y) \qquad (15)$$

$$He_{2r+1}^{(1)}(-|s|, u) = (-1)^{r+1} \sqrt{\frac{2}{\pi}} \sum_{m=0}^{\infty} \frac{Do_{2m+1}}{Do_1} \left[ I_m(x) K_{m+1}(y) - I_{m+1}(x) K_m(y) \right] \qquad (16)$$

$$Ho_{2r}^{(1)}(-|s|, u) = i(-1)^{r} \sqrt{\frac{2}{\pi}} \sum_{m=0}^{\infty} \frac{Do_{2m}}{Do_2} \left[ I_{m+1}(x) K_{m-1}(y) - I_{m-1}(x) K_{m+1}(y) \right] \qquad (17)$$

$$Ho_{2r+1}^{(1)}(-|s|, u) = (-1)^{r+1} \sqrt{\frac{2}{\pi}} \sum_{m=0}^{\infty} \frac{De_{2m+1}}{De_1} \left[ I_{m+1}(x) K_m(y) + I_m(x) K_{m+1}(y) \right] \qquad (18)$$

where $y = \frac{\sqrt{|s|}}{2} e^u$, $x = \frac{\sqrt{|s|}}{2} e^{-u}$, and $I_k(x)$ and $K_k(y)$ are the modified Bessel functions. Except when $s = 0$, the $He_n^{(1)}$ and $Ho_n^{(1)}$ functions are finite at $u = 0$.

For even integer $n$ and for negative parameter $s$, $He_n^{(1)}(-|s|, u)$ and $Ho_n^{(1)}(-|s|, u)$ are imaginary in form. It is sometimes more convenient to use real solutions of Equation (2) defined in a manner analogous to the manner of definition of the modified Bessel function of the second kind, as follows:

$$Ke_n(|s|, u) = i^{n+1} \sqrt{\frac{\pi}{2}} He_n^{(1)}(-|s|, u) \qquad (19)$$

$$Ko_n(|s|, u) = i^{n+1} \sqrt{\frac{\pi}{2}} Ho_n^{(1)}(-|s|, u) \qquad (20)$$

These relationships are consistent with the notation in Reference 4, page xxxiii. The $Ke_n$ and $Ko_n$ functions and derivatives may easily be calculated from the data given for the $He_n^{(1)}(-|s|, u)$ and $Ho_n^{(1)}(-|s|, u)$ functions and their derivatives in the tables of Reference 1 and this report.

The Mathieu Equations (1) and (2) may be obtained from the two dimensional scalar wave equation expressed in elliptical cylinder coordinates. The quantity s may be expressed as $k^2 d_o^2$, where k is a constant and $d_o$ is the semi-focal distance of the ellipse. As $d_o$ and s go to zero, the ellipse becomes a circle, and all of the radial Mathieu functions reduce to a constant times the Bessel function of the same kind with argument $(k\rho)$, where $\rho$ is the radial variable of the circular cylinder coordinate system. For example, $Je_n(s, u) \rightarrow \sqrt{\frac{\pi}{2}} \, J_n(k\rho)$ and $Ne_n(s, u) \rightarrow \sqrt{\frac{\pi}{2}} \, Y_n(k\rho)$ as $s \rightarrow o$, $u \neq o$. Also, $\cosh u = \frac{a}{d_o}$ where a is the length of the semi-major axis, so that as s and $d_o \rightarrow o$, u tends to infinity.

First Derivatives of the Mathieu Functions

The derivatives of the periodic Mathieu functions with respect to the angular variable v may be expressed as follows:

$$Se'_{2r}(s, v) = - \sum_{k=0}^{\infty} 2k \, De_{2k} \sin 2kv \tag{21}$$

$$Se'_{2r+1}(s, v) = - \sum_{k=0}^{\infty} (2k+1) \, De_{2k+1} \sin(2k+1)v \tag{22}$$

$$So'_{2r}(s, v) = \sum_{k=1}^{\infty} 2k \, Do_{2k} \cos 2kv \tag{23}$$

$$So'_{2r+1}(s, v) = \sum_{k=0}^{\infty} (2k+1) Do_{2k+1} \cos (2k+1) v \tag{24}$$

The derivatives of $Se_n$ and $So_n$ are absolutely and uniformly convergent for all real v (see discussion in Reference 5, page 38, on the derivatives of the analogous Mathieu functions $ce_n$ and $se_n$).

Equations for $Se_n$ and $So_n$ and their derivatives for negative parameter s may be simply obtained by the use of the following relationships (Reference 4, page xxxii):

$$De_{2k}(-|s|) = \frac{(-1)^k De_{2k}(|s|)}{Se_{2r}(|s|, \pi/2)} \tag{25}$$

$$De_{2k+1}(-|s|) = \frac{(-1)^k Do_{2k+1}(|s|)}{So_{2r+1}(|s|, \pi/2)} \tag{26}$$

$$Do_{2k+1}(-|s|) = \frac{(-1)^{k+1} De_{2k+1}(|s|)}{Se'_{2r+1}(|s|, \pi/2)} \tag{27}$$

$$Do_{2k}(-|s|) = \frac{(-1)^k Do_{2k}(|s|)}{So'_{2r}(|s|, \pi/2)} \tag{28}$$

Table I and Figures 1 through 12 give data on $Se'_n$ and $So'_n$ for n = 0, 1, 2 and various positive and negative real values of s; Table V and Figures 32 and 33 contain data on $Se_2(-|s|, v)$ and $So_2(-|s|, v)$.

The derivatives of the radial Mathieu functions of the first kind with respect to the radial variable u are given by these equations:

$$Je'_{2r+p}(s, u) = \frac{1}{2}\sqrt{\frac{\pi s}{2}}(-1)^r \sinh u \sum_{k=0}^{\infty}(-1)^k De_{2k+p}\left[J_{2k+p-1}(x) - J_{2k+p+1}(x)\right] \quad (29)$$

$$Jo'_{2r+p}(s, u) = \frac{Jo_{2r+p}}{\sinh u \cosh u} + \frac{1}{2}\sqrt{\frac{\pi s}{2}}(-1)^r \sinh u \tanh \cdot$$

$$\cdot \sum_{k=0}^{\infty}(-1)^k (2k+p) Do_{2k+p}\left[J_{2k+p-1}(x) - J_{2k+p+1}(x)\right] \quad (30)$$

where $x = \sqrt{s}\cosh u$, and $p = 0$ or $1$. In addition, at $u = 0$, $Je'_n(s, u) = 0$ and $Jo''_n(s, u) = 0$.

Equations (29) and (30) may be shown to be absolutely and uniformly convergent in any finite region of the u-plane. For a discussion of convergence of similar functions, see Reference 5, pages 168 and 169. Data on $Je'_n$ and $Jo'_n$ may be obtained from Table II and Figures 13 through 19 for $n = 0$, 1, and 2, $0 \leq u < 2$, and various positive values of $s$. Values for the derivative of $Jo_n$ at $u = 0$ were obtained with higher accuracy through the use of joining factors and relationships described in Reference 4 (page xxiii and the tables of joining factors).

Values of the derivatives with respect to u of the radial Mathieu functions of the second kind were computed from the following expressions:

$$Ne'_{2r}(s, u) = (-1)^r \frac{\sqrt{\pi/2}}{De_o}\sum_{k=0}^{\infty}(-1)^k De_{2k}\left[xY_k(y) J_{k+1}(x) - yY_{k+1}(y) J_k(x)\right] \quad (31)$$

$$Ne'_{2r+1}(s, u) = (-1)^r \frac{\sqrt{\frac{\pi}{2}}}{De_1} \sum_{k=0}^{\infty} (-1)^k De_{2k+1} \Big\{ x\Big[ Y_{k+1}(y) J_{k+1}(x) + Y_k(y) J_{k+2}(x) \Big]$$

$$- y\Big[ Y_{k+1}(y) J_{k+1}(x) + Y_{k+2}(y) J_k(x) \Big] + \Big[ Y_{k+1}(y) J_k(x) - Y_k(y) J_{k+1}(x) \Big] \Big\} \quad (32)$$

$$No'_{2r}(s, u) = (-1)^r \frac{\sqrt{\frac{\pi}{2}}}{Do_2} \sum_{k=1}^{\infty} (-1)^k Do_{2k} \Big\{ x\Big[ Y_{k+1}(y) J_k(x) + Y_{k-1}(y) J_k(x) \Big]$$

$$+ y\Big[ Y_k(y) J_{k+1}(x) + Y_k(y) J_{k-1}(x) \Big] - 2k\Big[ Y_{k+1}(y) J_{k-1}(x) + Y_{k-1}(y) J_{k+1}(x) \Big] \Big\} \quad (33)$$

$$No'_{2r+1}(s, u) = (-1)^r \frac{\sqrt{\frac{\pi}{2}}}{Do_2} \sum_{k=0}^{\infty} (-1)^k Do_{2k+1} \Big\{ x\Big[ Y_{k+1}(y) J_{k+1}(x) - Y_k(y) J_{k+2}(x) \Big]$$

$$+ y\Big[ Y_{k+1}(y) J_{k+1}(x) - Y_{k+2}(y) J_k(x) \Big] + \Big[ Y_{k+1}(y) J_k(x) + Y_k(y) J_{k+1}(x) \Big] \Big\} \quad (34)$$

where

$$y = \frac{\sqrt{s}}{2} e^u, \quad x = \frac{\sqrt{s}}{2} e^{-u} .$$

By various algebraic manipulations and the use of the recursion formulas of the Bessel functions, Equations (31) through (34) may be altered somewhat to suit computational needs. However, no further simplification of the series has been obvious. Simpler expressions for $Ne_n$ and $No_n$ analogous to the expansions used for $Je_n$ and $Jo_n$ are

$$Ne_{2r+p}(s, u) = \sqrt{\frac{\pi}{2}} (-1)^r \sum_{k=0}^{\infty} De_{2k+p} (-1)^k Y_{2k+p}(\sqrt{s} \cosh u) \tag{35}$$

$$No_{2r+p}(s, u) = \sqrt{\frac{\pi}{2}} (-1)^r \tanh u \sum_{k=0}^{\infty} (-1)^k (2k+p) Do_{2k+p} Y_{2k+p}(\sqrt{s} \cosh u). \tag{36}$$

These expansions, however, have certain restrictions (see Reference 4, page xxi)  as well as  very slow convergence for $0 \le u < 1$.  In addition, the derivatives of these expressions are not uniformly convergent at $u = 0$ (Reference 5, page 169).  Equations (35) and (36) are of use, however, in observing the behavior of $Ne_n$ and $No_n$ when $s \rightarrow 0$ (see earlier comments).  By analysis similar to that in Reference 5, page 257, Equations (31) through (34) may easily be shown to be absolutely and uniformly convergent in any finite region of the u-plane, including the origin.  The data for $Ne'_n(s, u)$ and $No'_n(s, u)$ for $n = 0, 1,$ and 2,  $0 \le u < 2$,  and various positive values of $s$ are contained in Table III and Figures 20 through 26.  More accurate values of the derivatives at $u = 0$ were obtained in the same way as those for $Jo'_n$. Additional expressions applicable at $u = 0$, which require only tables of the coefficients $De_m$ and $Do_m$, may be derived from the Bessel function product relationships and are listed below:

$$No_{2r}(s, o) = (-1)^{r+1} \frac{4\sqrt{2/\pi}}{s\, Do_2} \sum_{k=1}^{\infty} (-1)^k 2k\, Do_{2k} \tag{37}$$

$$No_{2r+1}(s, o) = (-1)^{r+1} \frac{2}{Do_1} \sqrt{\frac{2}{\pi s}} \sum_{k=0}^{\infty} (-1)^k Do_{2k+1} \tag{38}$$

$$\left[ Ne'_{2r}(s, u) \right]_{u=0} = (-1)^r \frac{1}{De_o} \sqrt{\frac{2}{\pi}} \sum_{k=0}^{\infty} (-1)^k De_{2k} \tag{39}$$

$$\left[Ne'_{2r+1}(s, u)\right]_{u=0} = (-1)^r \frac{2}{De_1} \sqrt{\frac{2}{\pi s}} \sum_{k=0}^{\infty} (-1)^k (2k+1) De_{2k+1} \qquad (40)$$

The data for the $Ne'_n$ and $No'_n$ functions are less accurate than for the $Je'_n$ and $Jo'_n$ functions. This occurs because of the cross-linked arguments of the Bessel function products and the difficulty encountered using tables of Bessel functions which carry arguments to only one or two decimal places. Fewer points where obtainable for the $Ne'_n$ and $No'_n$ functions near their maxima, minima, and zero crossings for the same reasons.

The derivatives with respect to the variable u of the radial Mathieu functions of the third kind, with the negative real parameter s, were obtained from the Bessel product series expansions of Equations (15) through (18). These derivatives are listed below:

$$He^{(1)'}_{2r}(-|s|, u) = i(-1)^r \frac{\sqrt{\frac{2}{\pi}}}{De_o} \sum_{m=0}^{\infty} De_{2m}\left[y\, I_m(x)\, K_{m-1}(y) + x\, I_{m-1}(x)\, K_m(y)\right] \qquad (41)$$

$$He^{(1)'}_{2r+1}(-|s|, u) = (-1)^{r+1} \frac{\sqrt{\frac{2}{\pi}}}{Do_1} \sum_{m=0}^{\infty} Do_{2m+1}\left\{y\left[I_{m+1}(x)\, K_{m-1}(y) - I_m(x)\, K_m(y)\right]\right.$$

$$\left. + x\left[I_m(x)\, K_m(y) - I_{m-1}(x)\, K_{m+1}(y)\right] - \left[I_m(x)\, K_{m+1}(y) + I_{m+1}(x)\, K_m(y)\right]\right\} \qquad (42)$$

$$Ho_{2r}^{(1)'}(-|s|, u) = i(-1)^{r+1} \frac{\sqrt{\frac{2}{\pi}}}{Do_2} \sum_{m=0}^{\infty} Do_{2m} \left\{ y \left[ I_{m+1}(x) K_{m-2}(y) - I_{m-1}(x) K_m(y) \right] \right.$$

$$+ x \left[ I_m(x) K_{m-1}(y) - I_{m-2}(x) K_{m+1}(y) \right] - 2 \left[ I_{m-1}(x) K_{m+1}(y) - I_{m+1}(x) K_{m-1}(y) \right] \right\}$$

$$(43)$$

$$Ho_{2r+1}^{(1)'}(-|s|, u) = (-1)^r \frac{\sqrt{\frac{2}{\pi}}}{De_1} \sum_{m=0}^{\infty} De_{2m+1} \left\{ y \left[ I_{m+1}(x) K_{m-1}(y) + I_m(x) K_m(y) \right] \right.$$

$$+ x \left[ I_m(x) K_m(y) + I_{m-1}(x) K_{m+1}(y) \right] - \left[ I_{m+1}(x) K_m(y) - I_m(x) K_{m+1}(y) \right] \right\} \quad (44)$$

where

$$y = \frac{\sqrt{|s|}}{2} e^u, \quad x = \frac{\sqrt{|s|}}{2} e^{-u} .$$

Data on these functions may be found in Table IV and Figures 27 through 31 for $0 \le u < 2$, $n = 0, 1,$ and 2, and various negative values of $s$.

Alternate expansions for $He_n^{(1)}$ and $Ho_n^{(1)}$ (and their derivatives) for negative $s$ in terms of the modified Bessel functions of the second kind may be derived which are analogous to Equations (35) and (36). However, these expressions converge slowly for small $u$ and are not uniformly convergent at $u = 0$ (see Reference 5, page 169). Their computational value is further limited by the fact that tables of $K_m(x)$ for $x > 5$ have not been available, and the $De_m$ and $Do_m$ coefficients are not tabulated for large $m$. These alternate expansions are of use, however, in examining $He_n^{(1)}$ and $Ho_n^{(1)}$ when $s \to 0$, since $u$ cannot be zero when $s = 0$.

Since

$$J_m(u) \sim I_m(u) \tag{45}$$

and

$$Y_m(u) \sim -\frac{2}{\pi} K_m(u) \tag{46}$$

when m is very large and much greater than $|u|$, Equations (41) through (44) may be shown to be absolutely and uniformly convergent for any finite region of the u-plane (see Reference 5, page 257).

Forms for finding $Ho_n^{(1)}(-|s|, u)$ at $u = 0$ are contained in Reference 1. In addition, equations for $He_n^{(1)'}(-|s|, u)$ at $u = 0$ have been obtained from the identities for products of Bessel functions and the fact that the $De_m$ and $Do_m$ coefficients are normalized as follows:

$$\sum_{k=0}^{\infty} De_{2k+p} = 1 , \qquad \sum_{k=0}^{\infty} (2k+p) Do_{2k+p} = 1 , \tag{47}$$

where $p = 0$ or $1$.

These equations are:

$$\left[ He_{2r}^{(1)'}(-|s|, u) \right]_{u=0} = i(-1)^r \frac{\sqrt{\frac{2}{\pi}}}{De_o} \tag{48}$$

$$\left[ He_{2r+1}^{(1)'}(-|s|, u) \right]_{u=0} = (-1)^r \frac{2}{Do_1} \sqrt{\frac{2}{\pi|s|}} \tag{49}$$

103

## Zeros of $Je_n(s, u)$, $Jo_n(s, u)$, $Je'_n(s, u)$, $Jo'_n(s, u)$

Tables VI, VII, and VIII contain the values of the variable u which give the first and second zeros of the radial Mathieu functions of the first kind and the first zeros of their derivatives. The cases where the functions or derivatives are zero at $u = 0$ are not counted as first zeros, since the condition $u = 0$ is usually associated with a trivial physical situation. The values of u were obtained numerically to three-decimal place accuracy by linear interpolation from values of the functions (or derivatives) calculated at arguments as close to the zero-crossings as the use of the tables of Bessel functions would permit. Additional zero values may be read (with less accuracy) from the graphs.

## General Remarks

In any set of tables the question of possible error is of great importance. The calculations were performed manually in the present instance, but much care has been taken to assure accuracy. The positions of the zero-crossings of the derivatives were cross-checked graphically with the positions of the maxima and minima of the curves of the original functions. Also, the positions and magnitudes of the maximum slopes of the functions have been compared with positions and magnitudes of maxima or minima of the derivatives. The original graphs prepared for this report and Reference 1 have horizontal and vertical scales approximately twice as large as the reduced copies used as figures and this permitted higher accuracies for graphical comparisons. The fact that the function and derivative curves are smoothly varying and mutually consistent reduces the possibility of the presence of any significant individual errors.

The values of the functions and derivatives at zero arguments were frequently calculated by more than one method, and in some cases these points have greater accuracy than that generally prevailing throughout the tables. As a further check, between five and ten per cent of the tabulated points were computed with higher precision than is shown in the tables. It is estimated that the tabulated values of the derivatives have a maximum error of plus or minus two units in the last decimal place, while the $Se_2(-|s|, v)$ and $So_2(-|s|, v)$ functions are estimated to have an error not greater than plus or minus one unit in the last place.

# REFERENCES

1. Wiltse, J. C., and King, M. J., Values of the Mathieu Functions, The Johns Hopkins University, Radiation Laboratory, Technical Report AF-53, August, 1958.

2. Chu, L. J., " Electromagnetic Waves in Elliptic Hollow Pipes of Metal, " Journal of Applied Physics, Vol. 9, p. 583, September, 1938.

3. Beattie, C. L., " Table of First 700 Zeros of Bessel Functions - $J_\ell(x)$ and $J'_\ell(x)$, " Bell System Technical Journal, Vol. XXXVII, No. 3, p. 689, May, 1958.

4. National Bureau of Standards, Tables Relating to Mathieu Functions, Columbia University Press, New York, 1951.

5. McLachlan, N. W., Theory and Application of Mathieu Functions, University Press, Oxford, 1951.

## TABLE I

### Derivatives with Respect to v of Periodic Mathieu Functions

$$Se_o'(s, v) = - \sum_{k=o}^{\infty} 2k\, De_{2k} \cdot \sin 2kv$$

| v | s = 1 | 3 | 6 | 9 | 15 | 25 | 40 |
|---|---|---|---|---|---|---|---|
| 0° | 0.0000 | 0.0000 | 0.0000 | 0.0000 | 0.0000 | 0.0000 | 0.0000 |
| 15° | 0.1339 | 0.4532 | 1.0389 | 1.7175 | 3.2783 | 6.3962 | 12.2904 |
| 30° | 0.2373 | 0.8426 | 2.0753 | 3.6824 | 8.0483 | 19.3223 | 48.9021 |
| 45° | 0.2828 | 1.0687 | 2.8817 | 5.5654 | 14.1701 | 42.2385 | 139.587 |
| 60° | 0.2527 | 1.0138 | 2.9706 | 6.1808 | 17.8956 | 63.5036 | 258.584 |
| 75° | 0.1492 | 0.6247 | 1.9375 | 4.2364 | 13.3336 | 52.8725 | 245.612 |
| 90° | 0.0000 | 0.0000 | 0.0000 | 0.0000 | 0.0000 | 0.0000 | 0.0000 |

$$Se_1'(s, v) = - \sum_{k=o}^{\infty} (2k+1)\, De_{2k+1}\, \sin(2k+1)v$$

| v | s = 1 | 3 | 6 | 9 | 15 | 25 | 40 |
|---|---|---|---|---|---|---|---|
| 0° | 0.0000 | 0.0000 | 0.0000 | 0.0000 | 0.0000 | 0.0000 | 0.0000 |
| 15° | -0.1984 | -0.0628 | +0.1844 | +0.4938 | +1.3161 | + 3.2249 | + 7.1081 |
| 30° | -0.4175 | -0.2225 | +0.1642 | +0.6965 | +2.3276 | + 7.1132 | +20.6290 |
| 45° | -0.6585 | -0.5306 | -0.2340 | +0.2429 | +2.0294 | + 8.9740 | +36.1957 |
| 60° | -0.8930 | -0.9454 | -1.0104 | -1.0337 | -0.7610 | + 2.1184 | +21.7142 |
| 75° | -1.0688 | -1.3192 | -1.8401 | -2.5998 | -5.1442 | -13.8678 | -44.6971 |
| 90° | -1.1346 | -1.4713 | -2.2044 | -3.3293 | -7.3981 | -23.2551 | -90.4954 |

$$Se'_2(s,v) = -\sum_{k=0}^{\infty} 2kDe_{2k}\sin 2kv$$

| v | s = 1 | 3 | 6 | 9 | 15 | 25 | 40 |
|---|---|---|---|---|---|---|---|
| 0° | 0.0000 | 0.0000 | 0.0000 | 0.0000 | 0.0000 | 0.0000 | 0.0000 |
| 15° | -0.8918 | -0.7064 | -0.4821 | -0.2832 | +0.1356 | +1.0928 | + 3.3781 |
| 30° | -1.5938 | -1.3548 | -1.0676 | -0.8132 | -0.2441 | +1.3316 | + 6.4880 |
| 45° | -1.9196 | -1.7848 | -1.6507 | -1.5651 | -1.4043 | -0.6983 | + 3.6842 |
| 60° | -1.7327 | -1.7506 | -1.8480 | -2.0330 | -2.6587 | -4.6452 | -10.5728 |
| 75° | -1.0307 | -1.1027 | -1.2663 | -1.5168 | -2.3619 | -5.5309 | -19.0947 |
| 90° | 0.0000 | 0.0000 | 0.0000 | 0.0000 | 0.0000 | 0.0000 | 0.0000 |

$$So'_1(s,v) = \sum_{k=0}^{\infty} (2k+1)Do_{2k+1}\cos(2k+1)v$$

| v | s = 1 | 3 | 6 | 9 | 15 | 25 | 40 |
|---|---|---|---|---|---|---|---|
| 0° | 1.0000 | 1.0000 | 1.0000 | 1.0000 | 1.0000 | 1.0000 | 1.0000 |
| 15° | 0.9906 | 1.0419 | 1.1235 | 1.2101 | 1.3968 | 1.7435 | 2.3433 |
| 30° | 0.9495 | 1.1300 | 1.4356 | 1.7841 | 2.6170 | 4.4569 | 8.5212 |
| 45° | 0.8458 | 1.1613 | 1.7393 | 2.4583 | 4.3984 | 9.5736 | 24.1974 |
| 60° | 0.6496 | 1.0073 | 1.7127 | 2.6612 | 5.5027 | 14.3593 | 44.8058 |
| 75° | 0.3564 | 0.5980 | 1.0996 | 1.8101 | 4.0910 | 11.9506 | 42.5561 |
| 90° | 0.0000 | 0.0000 | 0.0000 | 0.0000 | 0.0000 | 0.0000 | 0.0000 |

$$So'_2(s,v) = \sum_{k=1}^{\infty} 2kDo_{2k}\cos 2kv$$

| v | s = 1 | 3 | 6 | 9 | 15 | 25 | 40 |
|---|---|---|---|---|---|---|---|
| 0° | +1.0000 | +1.0000 | +1.0000 | +1.0000 | +1.0000 | +1.0000 | + 1.0000 |
| 15° | +0.8815 | +0.9137 | +0.9654 | +1.0210 | +1.1440 | +1.3827 | + 1.8132 |
| 30° | +0.5437 | +0.6344 | +0.7888 | +0.9656 | +1.3926 | +2.3575 | - 4.5386 |
| 45° | +0.0434 | +0.1417 | +0.3212 | +0.5457 | +1.1619 | +2.8612 | + 7.8529 |
| 60° | -0.4993 | -0.4925 | -0.4662 | -0.4141 | -0.1976 | +0.7194 | + 4.7127 |
| 75° | -0.9249 | -1.0554 | -1.2866 | -1.5666 | -2.3060 | -4.2424 | - 9.6445 |
| 90° | -1.0869 | -1.2837 | -1.6453 | -2.1028 | -3.3901 | -7.1565 | -19.5458 |

$$Se'_o(-|s|, v) = \frac{-1}{Se_o(|s|, \pi/2)} \sum_{k=0}^{\infty} (-1)^k 2k \, De_{2k} \sin 2kv$$

| v | $|s|=1$ | 3 | 6 | 9 | 15 |
|---|---|---|---|---|---|
| $0°$ | 0.0000 | 0.0000 | 0.0000 | 0.0000 | 0.0000 |
| $15°$ | -0.1163 | -0.3014 | -0.4938 | -0.6287 | -0.8204 |
| $30°$ | -0.1969 | -0.4891 | -0.7571 | -0.9173 | -1.1011 |
| $45°$ | -0.2204 | -0.5156 | -0.7344 | -0.8260 | -0.8719 |
| $60°$ | -0.1850 | -0.4065 | -0.5289 | -0.5465 | -0.4952 |
| $75°$ | -0.1044 | -0.2186 | -0.2648 | -0.2549 | -0.2017 |
| $90°$ | 0.0000 | 0.0000 | 0.0000 | 0.0000 | 0.0000 |

$$Se'_1(-|s|, v) = \frac{-1}{So_1(|s|, \pi/2)} \sum_{k=0}^{\infty} (-1)^k (2k+1) Do_{2k+1} \sin(2k+1)v$$

| v | $|s|=1$ | 3 | 6 | 9 | 15 |
|---|---|---|---|---|---|
| $0°$ | 0.0000 | 0.0000 | 0.0000 | 0.0000 | 0.0000 |
| $15°$ | -0.3149 | -0.4159 | -0.5446 | -0.6527 | -0.8267 |
| $30°$ | -0.5741 | -0.7006 | -0.8484 | -0.9596 | -1.1120 |
| $45°$ | -0.7474 | -0.8077 | -0.8616 | -0.8865 | -0.8888 |
| $60°$ | -0.8390 | -0.7860 | -0.7111 | -0.6433 | -0.5288 |
| $75°$ | -0.8753 | -0.7246 | -0.5565 | -0.4364 | -0.2823 |
| $90°$ | -0.8837 | -0.6955 | -0.4953 | -0.3606 | -0.2021 |

$$Se'_2(-|s|, v) = \frac{-1}{Se_2(|s|, \pi/2)} \sum_{k=0}^{\infty} (-1)^k 2k \, De_{2k} \sin 2kv$$

| v | $|s|=1$ | 3 | 6 | 9 | 15 |
|---|---|---|---|---|---|
| $0°$ | 0.0000 | 0.0000 | 0.0000 | 0.0000 | 0.0000 |
| $15°$ | -1.0512 | -1.3860 | -1.8242 | -2.2658 | -3.0726 |
| $30°$ | -1.7672 | -2.2003 | -2.6621 | -3.0368 | -3.4587 |
| $45°$ | -1.9578 | -2.2433 | -2.3780 | -2.3379 | -1.8268 |
| $60°$ | -1.6255 | -1.7029 | -1.5380 | -1.2147 | -0.3175 |
| $75°$ | -0.9095 | -0.8879 | -0.6946 | -0.4230 | +0.1764 |
| $90°$ | 0.0000 | 0.0000 | 0.0000 | 0.0000 | 0.0000 |

$$So'_1(-|s|, v) = \frac{1}{Se'_1(|s|, \pi/2)} \sum_{k=o}^{\infty} (-1)^{k+1}(2k+1) De_{2k+1} \cos(2k+1)v$$

| v | $\|s\|=1$ | 3 | 6 | 9 | 15 |
|------|---------|---------|---------|---------|---------|
| 0° | 1.0000 | +1.0000 | +1.0000 | +1.0000 | +1.0000 |
| 15° | 0.9421 | +0.8966 | +0.8348 | +0.7809 | +0.6953 |
| 30° | 0.7871 | +0.6426 | +0.4584 | +0.3105 | +0.1029 |
| 45° | 0.5804 | +0.3606 | +0.1062 | -0.0730 | -0.2743 |
| 60° | 0.3680 | +0.1510 | -0.0744 | -0.2092 | -0.3146 |
| 75° | 0.1749 | +0.0427 | -0.0836 | -0.1483 | -0.1779 |
| 90° | 0.0000 | 0.0000 | 0.0000 | 0.0000 | 0.0000 |

$$So'_2(-|s|, v) = \frac{1}{So'_2(|s|, \pi/2)} \sum_{k=1}^{\infty} (-1)^k 2k\, Do_{2k} \cos 2kv$$

| v | $\|s\|=1$ | 3 | 6 | 9 | 15 |
|------|---------|---------|---------|---------|---------|
| 0° | +1.0000 | +1.0000 | +1.0000 | +1.0000 | +1.0000 |
| 15° | +0.8510 | +0.8222 | +0.7820 | +0.7450 | +0.6802 |
| 30° | +0.4594 | +0.3837 | +0.3112 | +0.1969 | +0.0583 |
| 45° | -0.0400 | -0.1104 | -0.1952 | -0.2595 | -0.3427 |
| 60° | -0.5002 | -0.4942 | -0.4794 | -0.4592 | -0.4108 |
| 75° | -0.8110 | -0.7118 | -0.5867 | -0.4855 | -0.3375 |
| 90° | -0.9201 | -0.7790 | -0.6078 | -0.4756 | -0.2950 |

# TABLE II

Derivatives with Respect to u of the
Radial Mathieu Functions of the First Kind

$$Je'_o (s,u)$$

$$x = \sqrt{s} \cosh u$$

s = 1

| u | $Je'_o(s,u)$ | x | | u | $Je'_o(s,u)$ | x |
|---|---|---|---|---|---|---|
| 0.0000 | 0.0000 | 1.00 | | 1.5668 | -1.569 | 2.50 |
| 0.4436 | -0.290 | 1.10 | | 1.6892 | -1.497 | 2.80 |
| 0.6224 | -0.445 | 1.20 | | 1.7628 | -1.361 | 3.00 |
| 0.8671 | -0.716 | 1.40 | | 1.8309 | -1.155 | 3.20 |
| 1.1232 | -1.074 | 1.70 | | 1.8946 | -0.886 | 3.40 |
| 1.3169 | -1.357 | 2.00 | | 1.9542 | -0.563 | 3.60 |
| | | | | 1.9827 | -0.384 | 3.70 |

| s = 3 | | | | s = 6 | | |
|---|---|---|---|---|---|---|
| u | $Je'_o(s,u)$ | x | | u | $Je'_o(s,u)$ | x |
| 0.0000 | 0.000 | | | 0.0000 | 0.000 | |
| 0.1441 | -0.234 | 1.75 | | 0.0205 | -0.060 | 2.45 |
| 0.2792 | -0.460 | 1.80 | | 0.4904 | -1.347 | 2.75 |
| 0.3670 | -0.612 | 1.85 | | 0.6585 | -1.632 | 3.00 |
| 0.5493 | -0.940 | 2.00 | | 0.7493 | -1.680 | 3.17 |
| 0.9100 | -1.523 | 2.50 | | 0.8959 | -1.510 | 3.50 |
| 1.0371 | -1.599 | 2.75 | | 1.0730 | -0.744 | 4.00 |
| 1.0437 | -1.604 | 2.76 | | 1.1711 | -0.031 | 4.33 |
| 1.0456 | -1.598 | 2.77 | | 1.4747 | -0.001 | 4.343 |
| 1.1462 | -1.539 | 3.00 | | 1.4750 | 0.001 | 4.344 |
| 1.1861 | -1.477 | 3.10 | | 1.3404 | 1.458 | 5.00 |
| 1.3289 | -1.032 | 3.50 | | 1.4483 | 2.207 | 5.50 |
| 1.4796 | -0.133 | 4.00 | | 1.5242 | 2.389 | 5.89 |
| 1.4972 | -0.001 | 4.064 | | 1.5445 | 2.367 | 6.00 |
| 1.4974 | 0.001 | 4.065 | | 1.6315 | 1.860 | 6.50 |
| 1.6086 | 0.918 | 4.50 | | 1.6857 | 1.181 | 6.84 |
| 1.7218 | 1.828 | 5.00 | | 1.7111 | 0.793 | 7.00 |
| 1.8228 | 2.324 | 5.50 | | 1.8524 | -1.866 | 8.00 |
| 1.8750 | 2.352 | 5.78 | | 1.9754 | -2.940 | 9.00 |
| 1.9974 | 1.537 | 6.50 | | 2.0 | -2.78 | 9.22 |

## $Je'_0 (s, u)$ (cont'd)

<table>
<tr><td colspan="3" align="center">s = 9</td><td colspan="3" align="center">s = 15</td></tr>
<tr><td>u</td><td>$Je'_0(s,u)$</td><td>x</td><td>u</td><td>$Je'_0(s,u)$</td><td>x</td></tr>
<tr><td>0.0000</td><td>0.000</td><td></td><td>0.0000</td><td>0.000</td><td></td></tr>
<tr><td>0.3045</td><td>-1.182</td><td>3.14</td><td>0.0602</td><td>-0.378</td><td>3.88</td></tr>
<tr><td>0.5697</td><td>-1.764</td><td>3.50</td><td>0.2554</td><td>-1.456</td><td>4.00</td></tr>
<tr><td>0.6020</td><td>-1.774</td><td>3.56</td><td>0.5616</td><td>-1.780</td><td>4.50</td></tr>
<tr><td>0.7954</td><td>-1.392</td><td>4.00</td><td>0.6386</td><td>-1.424</td><td>4.69</td></tr>
<tr><td>0.9888</td><td>-0.112</td><td>4.59</td><td>0.7456</td><td>-0.601</td><td>5.00</td></tr>
<tr><td>1.0987</td><td>0.921</td><td>5.00</td><td>0.8058</td><td>0.004</td><td>5.20</td></tr>
<tr><td>1.2149</td><td>1.948</td><td>5.50</td><td>0.9602</td><td>1.672</td><td>5.80</td></tr>
<tr><td>1.3170</td><td>2.413</td><td>6.00</td><td>1.0052</td><td>2.066</td><td>6.00</td></tr>
<tr><td>1.4083</td><td>2.159</td><td>6.50</td><td>1.1073</td><td>2.494</td><td>6.50</td></tr>
<tr><td>1.4910</td><td>1.239</td><td>7.00</td><td>1.1979</td><td>2.060</td><td>7.00</td></tr>
<tr><td>1.5697</td><td>-0.163</td><td>7.52</td><td>1.2605</td><td>1.237</td><td>7.38</td></tr>
<tr><td>1.6368</td><td>-1.504</td><td>8.00</td><td>1.3540</td><td>-0.588</td><td>8.00</td></tr>
<tr><td>1.7107</td><td>-2.668</td><td>8.57</td><td>1.4602</td><td>-2.583</td><td>8.79</td></tr>
<tr><td>1.7627</td><td>-2.979</td><td>9.00</td><td>1.5131</td><td>-3.008</td><td>9.22</td></tr>
<tr><td>1.7883</td><td>-2.911</td><td>9.22</td><td>1.6019</td><td>-2.183</td><td>10.00</td></tr>
<tr><td>1.8738</td><td>-1.502</td><td>10.00</td><td>1.6505</td><td>-0.894</td><td>10.46</td></tr>
<tr><td>1.8997</td><td>-0.758</td><td>10.25</td><td>1.7045</td><td>0.931</td><td>11.00</td></tr>
<tr><td>1.9733</td><td>1.670</td><td>11.00</td><td>1.7969</td><td>3.348</td><td>12.00</td></tr>
<tr><td></td><td></td><td></td><td>1.8811</td><td>2.507</td><td>13.00</td></tr>
<tr><td></td><td></td><td></td><td>1.9585</td><td>-0.953</td><td>14.00</td></tr>
</table>

## $Je'_1 (s, u)$

$$x = \sqrt{s} \cosh u$$

<table>
<tr><td colspan="3" align="center">s = 1</td><td colspan="3" align="center">s = 3</td></tr>
<tr><td>u</td><td>$Je'_1(s,u)$</td><td>x</td><td>u</td><td>$Je'_1(s,u)$</td><td>x</td></tr>
<tr><td>0.0000</td><td>0.000</td><td></td><td>0.0000</td><td>0.000</td><td></td></tr>
<tr><td>0.4436</td><td>0.174</td><td>1.10</td><td>0.1441</td><td>0.018</td><td>1.75</td></tr>
<tr><td>0.6224</td><td>0.222</td><td>1.20</td><td>0.3670</td><td>0.010</td><td>1.85</td></tr>
<tr><td>0.8671</td><td>0.232</td><td>1.40</td><td>0.5493</td><td>-0.067</td><td>2.00</td></tr>
<tr><td>1.1232</td><td>0.111</td><td>1.70</td><td>0.9100</td><td>-0.572</td><td>2.50</td></tr>
<tr><td>1.2269</td><td>0.001</td><td>1.852</td><td>1.1462</td><td>-1.209</td><td>3.00</td></tr>
<tr><td>1.3169</td><td>-0.133</td><td>2.00</td><td>1.3289</td><td>-1.719</td><td>3.50</td></tr>
<tr><td>1.5668</td><td>-0.716</td><td>2.50</td><td>1.4796</td><td>-1.887</td><td>4.00</td></tr>
<tr><td>1.6892</td><td>-1.102</td><td>2.80</td><td>1.6086</td><td>-1.592</td><td>4.50</td></tr>
<tr><td>1.7628</td><td>-1.345</td><td>3.00</td><td>1.7218</td><td>-0.844</td><td>5.00</td></tr>
<tr><td>1.8309</td><td>-1.561</td><td>3.20</td><td>1.8228</td><td>0.212</td><td>5.50</td></tr>
<tr><td>1.8946</td><td>-1.735</td><td>3.40</td><td>1.9974</td><td>2.213</td><td>6.50</td></tr>
<tr><td>1.9542</td><td>-1.856</td><td>3.60</td><td></td><td></td><td></td></tr>
<tr><td>1.9827</td><td>-1.894</td><td>3.70</td><td></td><td></td><td></td></tr>
</table>

$$Je'_1(s,u) \quad (cont'd)$$

|  | s = 6 |  |  |  | s = 9 |  |
|---|---|---|---|---|---|---|
| u | $Je'_1(s,u)$ | x |  | u | $Je'_1(s,u)$ | x |
| 0.0000 | 0.000 |  |  | 0.0000 | 0.000 |  |
| 0.0205 | -0.015 | 2.45 |  | 0.3045 | -0.570 | 3.14 |
| 0.4904 | -0.548 | 2.75 |  | 0.5697 | -1.236 | 3.50 |
| 0.6585 | -0.913 | 3.00 |  | 0.7954 | -1.768 | 4.00 |
| 0.8959 | -1.543 | 3.50 |  | 0.8904 | -1.846 | 4.27 |
| 1.0730 | -1.851 | 4.00 |  | 1.0987 | -1.247 | 5.00 |
| 1.1711 | -1.799 | 4.33 |  | 1.2149 | -0.275 | 5.50 |
| 1.3404 | -1.025 | 5.00 |  | 1.2407 | -0.001 | 5.621 |
| 1.4483 | -0.003 | 5.50 |  | 1.2409 | 0.002 | 5.622 |
| 1.4485 | -0.001 | 5.501 |  | 1.3170 | 0.884 | 6.00 |
| 1.4487 | 0.001 | 5.502 |  | 1.4083 | 1.917 | 6.50 |
| 1.6315 | 2.086 | 6.50 |  | 1.4910 | 2.525 | 7.00 |
| 1.7111 | 2.579 | 7.00 |  | 1.5697 | 2.492 | 7.52 |
| 1.7744 | 2.502 | 7.43 |  | 1.6368 | 1.825 | 8.00 |
| 1.8524 | 1.667 | 8.00 |  | 1.7627 | -0.832 | 9.00 |
| 1.9305 | 0.067 | 8.62 |  | 1.7883 | -1.451 | 9.22 |
| 2.0 | -1.634 | 9.22 |  | 1.8738 | -2.994 | 10.00 |
|  |  |  |  | 1.8997 | -3.153 | 10.25 |
|  |  |  |  | 1.9733 | -2.440 | 11.00 |

$$s = 15$$

| u | $Je'_1(s,u)$ | x |  | u | $Je'_1(s,u)$ | x |
|---|---|---|---|---|---|---|
| 0.0000 | 0.000 |  |  | 1.1979 | 2.302 | 7.00 |
| 0.0602 | -0.215 | 3.88 |  | 1.2605 | 2.596 | 7.38 |
| 0.2554 | -0.913 | 4.00 |  | 1.3540 | 2.187 | 8.00 |
| 0.5616 | -1.809 | 4.50 |  | 1.4602 | 0.323 | 8.79 |
| 0.6386 | -1.875 | 4.69 |  | 1.5131 | -0.960 | 9.22 |
| 0.7456 | -1.739 | 5.00 |  | 1.6019 | -2.796 | 10.00 |
| 0.9889 | -0.002 | 5.926 |  | 1.6505 | -3.163 | 10.46 |
| 0.9891 | 0.001 | 5.927 |  | 1.7045 | -2.713 | 11.00 |
| 1.0052 | 0.183 | 6.00 |  | 1.7969 | 0.109 | 12.00 |
| 1.1073 | 1.401 | 6.50 |  | 1.8811 | 3.073 | 13.00 |
|  |  |  |  | 1.9585 | 3.262 | 14.00 |

$$Je'_1 (s, u) \quad (cont'd)$$

| | s = 25 | | | | s = 40 | |
| u | $Je'_1(s,u)$ | x | | u | $Je'_1(s,u)$ | x |
|---|---|---|---|---|---|---|
| 0.0000 | 0.000 | | | 0.0000 | 0.000 | |
| 0.4436 | -2.021 | 5.50 | | 0.0415 | -0.430 | 6.33 |
| 0.6224 | -1.442 | 6.00 | | 0.2350 | -2.024 | 6.50 |
| 0.7141 | -0.610 | 6.33 | | 0.3290 | -2.265 | 6.67 |
| 0.7564 | -0.117 | 6.50 | | 0.4581 | -1.716 | 7.00 |
| 0.7658 | -0.002 | 6.539 | | 0.6006 | 0.069 | 7.50 |
| 0.7660 | 0.001 | 6.540 | | 0.7127 | 1.791 | 8.00 |
| 0.8671 | 1.320 | 7.00 | | 0.8073 | 2.695 | 8.50 |
| 1.0470 | 2.640 | 8.00 | | 0.9237 | 2.088 | 9.22 |
| 1.1929 | 0.853 | 9.00 | | 1.0317 | -0.345 | 10.00 |
| 1.2218 | 0.180 | 9.22 | | 1.1513 | -3.041 | 11.00 |
| 1.3170 | -2.145 | 10.00 | | 1.2556 | -2.335 | 12.00 |
| 1.4254 | -3.110 | 11.00 | | 1.3484 | 1.107 | 13.00 |
| 1.5221 | -0.847 | 12.00 | | 1.4323 | 3.600 | 14.00 |
| 1.6094 | 2.453 | 13.00 | | 1.5090 | 2.405 | 15.00 |
| 1.6892 | 3.551 | 14.00 | | 1.5797 | -1.410 | 16.00 |
| 1.7628 | 1.100 | 15.00 | | 1.6454 | -4.012 | 17.00 |
| 1.8310 | -2.640 | 16.00 | | 1.7067 | -2.660 | 18.00 |
| 1.8946 | -4.009 | 17.00 | | 1.8185 | 4.344 | 20.00 |
| 1.9542 | -1.515 | 18.00 | | 1.9184 | -1.121 | 22.00 |

$$Je'_2 (s, u)$$

$$x = \sqrt{s} \ \cosh u$$

s = 1

| u | $Je'_2(s,u)$ | x | | u | $Je'_2(s,u)$ | x |
|---|---|---|---|---|---|---|
| 0.0000 | 0.0000 | | | 1.5668 | 0.4775 | 2.50 |
| 0.4436 | 0.1390 | 1.10 | | 1.6892 | 0.3036 | 2.80 |
| 0.6224 | 0.2108 | 1.20 | | 1.7628 | 0.1322 | 3.00 |
| 0.8671 | 0.3301 | 1.40 | | 1.8309 | -0.0791 | 3.20 |
| 1.1232 | 0.4683 | 1.70 | | 1.8946 | -0.3238 | 3.40 |
| 1.3169 | 0.5440 | 2.00 | | 1.9542 | -0.5930 | 3.60 |
| | | | | 1.9827 | -0.7336 | 3.70 |

$$Je'_2(s,u) \quad (cont'd)$$

| | s = 3 | | | s = 6 | |
|---|---|---|---|---|---|
| u | $Je'_2(s,u)$ | x | u | $Je'_2(s,u)$ | x |
| 0.0000 | 0.000 | | 0.0000 | 0.000 | |
| 0.1441 | 0.099 | 1.75 | 0.0205 | 0.017 | 2.45 |
| 0.2792 | 0.191 | 1.80 | 0.4904 | 0.305 | 2.75 |
| 0.3670 | 0.252 | 1.85 | 0.6585 | 0.256 | 3.00 |
| 0.5493 | 0.371 | 2.00 | 0.8332 | 0.001 | 3.35 |
| 0.6969 | 0.449 | 2.17 | 0.8959 | -0.160 | 3.50 |
| 0.9100 | 0.476 | 2.50 | 1.0730 | -0.829 | 4.00 |
| 1.0371 | 0.395 | 2.75 | 1.1711 | -1.305 | 4.33 |
| 1.1462 | 0.231 | 3.00 | 1.3404 | -2.033 | 5.00 |
| 1.1861 | 0.143 | 3.10 | 1.4483 | -2.131 | 5.50 |
| 1.3289 | -0.309 | 3.50 | 1.5242 | -1.860 | 5.89 |
| 1.4796 | -1.014 | 4.00 | 1.6315 | -0.855 | 6.50 |
| 1.6086 | -1.687 | 4.50 | 1.7111 | 0.317 | 7.00 |
| 1.7218 | -2.113 | 5.00 | 1.8524 | 2.447 | 8.00 |
| 1.8228 | -2.119 | 5.50 | 1.9305 | 2.841 | 8.62 |
| 1.9974 | -0.700 | 6.50 | 2.0 | 2.230 | 9.22 |

| | s = 9 | | | s = 15 | |
|---|---|---|---|---|---|
| u | $Je'_2(s,u)$ | x | u | $Je'_2(s,u)$ | x |
| 0.0000 | 0.000 | | 0.0000 | 0.000 | |
| 0.3045 | 0.120 | 3.14 | 0.0029 | -0.002 | 3.873 |
| 0.5697 | -0.063 | 3.50 | 0.0602 | -0.040 | 3.88 |
| 0.7953 | -0.661 | 4.00 | 0.2554 | -0.230 | 4.00 |
| 0.9888 | -1.495 | 4.59 | 0.5616 | -0.976 | 4.50 |
| 1.0987 | -1.936 | 5.00 | 0.6386 | -1.262 | 4.69 |
| 1.2149 | -2.119 | 5.50 | 0.7456 | -1.673 | 5.00 |
| 1.3170 | -1.798 | 6.00 | 0.9602 | -2.034 | 5.80 |
| 1.4083 | -0.985 | 6.50 | 1.0052 | -1.915 | 6.00 |
| 1.4910 | 0.162 | 7.00 | 1.1073 | -1.245 | 6.50 |
| 1.6368 | 2.355 | 8.00 | 1.1979 | -0.164 | 7.00 |
| 1.7019 | 2.819 | 8.50 | 1.2094 | 0.002 | 7.068 |
| 1.7627 | 2.612 | 9.00 | 1.2605 | 0.778 | 7.38 |
| 1.7883 | 2.303 | 9.22 | 1.3540 | 2.142 | 8.00 |
| 1.8738 | 0.379 | 10.00 | 1.4602 | 2.798 | 8.79 |
| 1.9733 | -2.482 | 11.00 | 1.5131 | 2.137 | 9.22 |
| | | | 1.6019 | 0.644 | 10.00 |
| | | | 1.6505 | -0.766 | 10.46 |
| | | | 1.7045 | -2.282 | 11.00 |
| | | | 1.7969 | -3.303 | 12.00 |
| | | | 1.8811 | -1.227 | 13.00 |
| | | | 1.9585 | 2.212 | 14.00 |

$$Jo'_1 (s, u)$$

$$x = \sqrt{s} \cosh u$$

| | s = 1 | | | | s = 3 | |
|---|---|---|---|---|---|---|
| u | $Jo'_1(s,u)$ | x | | u | $Jo'_1(s,u)$ | x |
| 0.0000 | 0.60805 | | | 0.0000 | 0.99767 | |
| 0.4436 | 0.6164 | 1.10 | | 0.1441 | 0.984 | 1.75 |
| 0.6224 | 0.6117 | 1.20 | | 0.2792 | 0.942 | 1.80 |
| 0.8671 | 0.5619 | 1.40 | | 0.3670 | 0.897 | 1.85 |
| 1.1232 | 0.3860 | 1.70 | | 0.5493 | 0.738 | 2.00 |
| 1.3169 | 0.0994 | 2.00 | | 0.9100 | 0.011 | 2.50 |
| 1.3620 | 0.0070 | 2.08 | | 0.9138 | -0.001 | 2.507 |
| 1.3674 | -0.0049 | 2.09 | | 0.9144 | -0.003 | 2.508 |
| 1.5668 | -0.5516 | 2.50 | | 0.9155 | -0.006 | 2.51 |
| 1.6892 | -0.9807 | 2.80 | | 1.0371 | -0.419 | 2.75 |
| 1.7628 | -1.2538 | 3.00 | | 1.1462 | -0.852 | 3.00 |
| 1.8309 | -1.4986 | 3.20 | | 1.1861 | -1.018 | 3.10 |
| 1.8946 | -1.7022 | 3.40 | | 1.3289 | -1.592 | 3.50 |
| 1.9542 | -1.8510 | 3.60 | | 1.4796 | -1.970 | 4.00 |
| 1.9827 | -1.9018 | 3.70 | | 1.6086 | -1.834 | 4.50 |
| | | | | 1.7218 | -1.080 | 5.00 |
| | | | | 1.8228 | -0.120 | 5.50 |
| | | | | 1.9974 | 2.073 | 6.50 |

| | s = 6 | | | | s = 9 | |
|---|---|---|---|---|---|---|
| u | $Jo'_1(s,u)$ | x | | u | $Jo'_1(s,u)$ | x |
| 0.0000 | 1.31774 | | | 0.0000 | 1.52555 | |
| 0.0205 | 1.317 | 2.45 | | 0.3974 | 0.758 | 3.24 |
| 0.4904 | 0.665 | 2.75 | | 0.5584 | 0.009 | 3.48 |
| 0.6585 | 0.078 | 3.00 | | 0.5640 | -0.021 | 3.49 |
| 0.6755 | 0.007 | 3.03 | | 0.5697 | -0.051 | 3.50 |
| 0.6811 | -0.017 | 3.04 | | 0.7953 | -1.344 | 4.00 |
| 0.8959 | -1.039 | 3.50 | | 0.9624 | -2.034 | 4.50 |
| 1.0730 | -1.819 | 4.00 | | 1.0987 | -1.970 | 5.00 |
| 1.1711 | -2.033 | 4.33 | | 1.2149 | -1.208 | 5.50 |
| 1.3404 | -1.615 | 5.00 | | 1.3170 | 0.008 | 6.00 |
| 1.4483 | -0.662 | 5.50 | | 1.4083 | 1.311 | 6.50 |
| 1.5661 | 0.870 | 6.12 | | 1.4910 | 2.314 | 7.00 |
| 1.6315 | 1.740 | 6.50 | | 1.5668 | 2.711 | 7.50 |
| 1.7111 | 2.514 | 7.00 | | 1.6368 | 2.358 | 8.00 |
| 1.7630 | 2.685 | 7.35 | | 1.7627 | -0.133 | 9.00 |
| 1.8524 | 2.079 | 8.00 | | 1.7883 | -0.813 | 9.22 |
| 1.9245 | 0.815 | 8.57 | | 1.8738 | -2.772 | 10.00 |
| | | | | 1.9248 | -3.224 | 10.50 |
| | | | | 1.9733 | -2.842 | 11.00 |

## Jo'₁ (s,u) (cont'd)

### s = 15

| u | Jo'₁(s,u) | x |
|---|---|---|
| 0.0000 | 1.80424 | |
| 0.0602 | 1.767 | 3.88 |
| 0.2554 | 1.149 | 4.00 |
| 0.4320 | 0.029 | 4.24 |
| 0.4360 | -0.001 | 4.247 |
| 0.4377 | -0.013 | 4.25 |
| 0.4490 | -0.098 | 4.27 |
| 0.5616 | -0.961 | 4.50 |
| 0.7456 | -2.072 | 5.00 |
| 0.8087 | -2.208 | 5.21 |
| 1.0052 | -1.202 | 6.00 |
| 1.1073 | 0.193 | 6.50 |
| 1.1979 | 1.596 | 7.00 |
| 1.3540 | 2.730 | 8.00 |
| 1.5131 | 0.088 | 9.22 |
| 1.6019 | -2.259 | 10.00 |
| 1.6546 | -3.127 | 10.50 |
| 1.7045 | -3.161 | 11.00 |
| 1.7969 | -0.814 | 12.00 |
| 1.8811 | 2.547 | 13.00 |
| 1.9585 | 3.598 | 14.00 |

### s = 25

| u | Jo'₁(s,u) | x |
|---|---|---|
| 0.0000 | 2.10321 | |
| 0.1413 | 1.69 | 5.05 |
| 0.3390 | 0.10 | 5.29 |
| 0.3424 | -0.01 | 5.296 |
| 0.3447 | -0.09 | 5.30 |
| 0.3503 | -0.15 | 5.31 |
| 0.3559 | -0.21 | 5.32 |
| 0.4436 | -1.14 | 5.50 |
| 0.6224 | -2.36 | 6.00 |
| 0.7141 | -2.21 | 6.33 |
| 0.7564 | -1.90 | 6.50 |
| 0.8671 | -0.44 | 7.00 |
| 1.0470 | 2.49 | 8.00 |
| 1.1929 | 2.20 | 9.00 |
| 1.2218 | 1.65 | 9.22 |
| 1.3170 | -0.94 | 10.00 |
| 1.4254 | -3.25 | 11.00 |
| 1.5221 | -2.05 | 12.00 |
| 1.6094 | 1.36 | 13.00 |
| 1.6892 | 3.66 | 14.00 |
| 1.7628 | 2.16 | 15.00 |
| 1.8310 | -1.73 | 16.00 |
| 1.9542 | -2.59 | 18 |

### s = 40

| u | Jo'₁(s,u) | x |
|---|---|---|
| 0.0000 | 2.40237 | |
| 0.0415 | 2.33 | 6.33 |
| 0.2350 | 0.43 | 6.50 |
| 0.2480 | 0.24 | 6.52 |
| 0.2542 | 0.14 | 6.53 |
| 0.2603 | 0.05 | 6.54 |
| 0.2662 | -0.04 | 6.55 |
| 0.3429 | -1.17 | 6.70 |
| 0.4581 | -2.41 | 7.00 |
| 0.6006 | -2.15 | 7.50 |
| 0.7127 | -0.33 | 8.00 |
| 0.8073 | 1.67 | 8.50 |

| u | Jo'₁(s,u) | x |
|---|---|---|
| 0.9237 | 2.98 | 9.22 |
| 1.0317 | 1.57 | 10.00 |
| 1.1513 | -2.08 | 11.00 |
| 1.2556 | -3.29 | 12.00 |
| 1.3484 | -1.40 | 13.00 |
| 1.4323 | 3.04 | 14.00 |
| 1.5090 | 3.41 | 15.00 |
| 1.5797 | -1.18 | 16.00 |
| 1.6454 | -3.59 | 17.00 |
| 1.7067 | -3.60 | 18.00 |
| 1.8185 | 3.94 | 20.00 |
| 1.8697 | 3.88 | 21.00 |
| 1.9184 | -0.11 | 22.00 |

$$Jo'_2(s,u)$$

$$x = \sqrt{s} \cosh u$$

| | s = 1 | | | | s = 3 | |
|---|---|---|---|---|---|---|
| u | $Jo'_2(s,u)$ | x | | u | $Jo'_2(s,u)$ | x |
| 0.0000 | 0.15033 | | | 0.0000 | 0.41624 | |
| 0.4436 | 0.2034 | 1.10 | | 0.1441 | 0.427 | 1.75 |
| 0.6224 | 0.2567 | 1.20 | | 0.2792 | 0.455 | 1.80 |
| 0.8671 | 0.3597 | 1.40 | | 0.3670 | 0.481 | 1.85 |
| 1.1232 | 0.4882 | 1.70 | | 0.5493 | 0.547 | 2.00 |
| 1.3169 | 0.5597 | 2.00 | | 0.9100 | 0.594 | 2.50 |
| 1.5668 | 0.4902 | 2.50 | | 1.0371 | 0.502 | 2.75 |
| 1.6892 | 0.3151 | 2.80 | | 1.1462 | 0.329 | 3.00 |
| 1.7628 | 0.1430 | 3.00 | | 1.1861 | 0.238 | 3.10 |
| 1.8076 | 0.0093 | 3.13 | | 1.2711 | -0.012 | 3.33 |
| 1.8110 | -0.0016 | 3.14 | | 1.2746 | -0.024 | 3.34 |
| 1.8309 | -0.0690 | 3.20 | | 1.3289 | -0.228 | 3.50 |
| 1.8633 | -0.1882 | 3.30 | | 1.4796 | -0.955 | 4.00 |
| 1.8946 | -0.3146 | 3.40 | | 1.6086 | -1.655 | 4.50 |
| 1.9542 | -0.5847 | 3.60 | | 1.7218 | -2.110 | 5.00 |
| 1.9827 | -0.7258 | 3.70 | | 1.8228 | -2.144 | 5.50 |
| | | | | 1.9974 | -0.753 | 6.50 |

| | s = 6 | | | | s = 9 | |
|---|---|---|---|---|---|---|
| u | $Jo'_2(s,u)$ | x | | u | $Jo'_2(s,u)$ | x |
| 0.0000 | 0.74276 | | | 0.0000 | 1.00180 | |
| 0.0205 | 0.743 | 2.45 | | 0.3045 | 0.940 | 3.14 |
| 0.4904 | 0.747 | 2.75 | | 0.5697 | 0.619 | 3.50 |
| 0.6585 | 0.642 | 3.00 | | 0.7644 | 0.005 | 3.92 |
| 0.8959 | 0.156 | 3.50 | | 0.7684 | -0.012 | 3.93 |
| 0.9386 | 0.008 | 3.61 | | 0.7953 | -0.132 | 4.00 |
| 0.9424 | -0.006 | 3.62 | | 0.9888 | -1.189 | 4.59 |
| 0.9897 | -0.196 | 3.75 | | 1.0987 | -1.805 | 5.00 |
| 1.0730 | -0.591 | 4.00 | | 1.2149 | -2.195 | 5.50 |
| 1.1711 | -1.128 | 4.33 | | 1.3170 | -2.041 | 6.00 |
| 1.3404 | -1.996 | 5.00 | | 1.4083 | -1.328 | 6.50 |
| 1.4483 | -2.195 | 5.50 | | 1.4910 | -0.195 | 7.00 |
| 1.5445 | -1.868 | 6.00 | | 1.5697 | 1.137 | 7.52 |
| 1.6315 | -1.037 | 6.50 | | 1.6368 | 2.190 | 8.00 |
| 1.7111 | 0.136 | 7.00 | | 1.7107 | 2.845 | 8.57 |
| 1.8524 | 2.375 | 8.00 | | 1.7627 | 2.750 | 9.00 |
| 1.9245 | 2.866 | 8.57 | | 1.7883 | 2.495 | 9.22 |
| 1.9754 | 2.648 | 9.00 | | 1.8738 | 0.669 | 10.00 |
| 2.0 | 2.335 | 9.22 | | 1.8997 | -0.104 | 10.25 |
| | | | | 1.9733 | -2.304 | 11.00 |

$Jo'_2(s, u)$ (cont'd)

$$s = 15$$

| u | $Jo'_2(s, u)$ | x | u | $Jo'_2(s, u)$ | x |
|---|---|---|---|---|---|
| 0.0000 | 1.38126 | | 1.1979 | -0.914 | 7.00 |
| 0.0602 | 1.369 | 3.88 | 1.2605 | 0.067 | 7.38 |
| 0.2554 | 1.157 | 4.00 | 1.3540 | 1.691 | 8.00 |
| 0.5616 | 0.067 | 4.50 | 1.4602 | 2.859 | 8.79 |
| 0.5703 | 0.020 | 4.52 | 1.5131 | 2.758 | 9.22 |
| 0.5745 | -0.003 | 4.53 | 1.6019 | 1.235 | 10.00 |
| 0.6386 | -0.383 | 4.69 | 1.6505 | -0.182 | 10.46 |
| 0.7456 | -1.086 | 5.00 | 1.7045 | -1.856 | 11.00 |
| 0.9602 | -2.208 | 5.80 | 1.7969 | -3.400 | 12.00 |
| 1.0052 | -2.249 | 6.00 | 1.8811 | -1.716 | 13.00 |
| 1.1073 | -1.878 | 6.50 | 1.9585 | 1.795 | 14.00 |

## TABLE III

### Derivatives with Respect to u of the Radial Mathieu Functions of the Second Kind

$$Ne_0'(s,u)$$

$$y = \frac{\sqrt{s}}{2} e^{u}, \quad x = \frac{\sqrt{s}}{2} e^{-u}$$

#### s = 1

| u | $Ne_0'(s,u)$ | y | x |
|---|---|---|---|
| 0.0000 | 0.89850 | | |
| 0.182 | 0.948 | 0.60 | .417 |
| 0.470 | 0.997 | 0.80 | .312 |
| 0.588 | 1.004 | 0.90 | .278 |
| 0.693 | 0.999 | 1.00 | .250 |
| 0.876 | 0.955 | 1.20 | .208 |
| 0.956 | 0.914 | 1.30 | .192 |
| 1.224 | 0.625 | 1.70 | .147 |
| 1.435 | 0.151 | 2.10 | .119 |
| 1.609 | -0.446 | 2.50 | .100 |
| 1.723 | -0.916 | 2.80 | .089 |
| 1.792 | -1.214 | 3.00 | .083 |
| 1.974 | -1.873 | 3.60 | .069 |

#### s = 3

| u | $Ne_0'(s,u)$ | y | x |
|---|---|---|---|
| 0.0000 | 1.09407 | | |
| 0.144 | 1.122 | 1.00 | .75 |
| 0.480 | 1.003 | 1.40 | .54 |
| 0.614 | 0.854 | 1.60 | .469 |
| 0.732 | 0.653 | 1.80 | .417 |
| 0.886 | 0.266 | 2.10 | .357 |
| 1.019 | -0.193 | 2.40 | .313 |
| 1.242 | -1.160 | 3.00 | .250 |
| 1.452 | -1.929 | 3.70 | .203 |
| 1.530 | -2.013 | 4.00 | .188 |
| 1.753 | -0.987 | 5.00 | .150 |
| 1.936 | 1.270 | 6.00 | .125 |

#### s = 6

| u | $Ne_0'(s,u)$ | y | x |
|---|---|---|---|
| 0.0000 | 1.34234 | | |
| 0.060 | 1.35 | 1.30 | 1.15 |
| 0.203 | 1.27 | 1.50 | 1.00 |
| 0.490 | 0.74 | 2.00 | 0.75 |
| 0.714 | -0.09 | 2.50 | 0.60 |
| 0.827 | -0.63 | 2.80 | 0.54 |
| 0.896 | -0.99 | 3.00 | 0.500 |
| 0.991 | -1.46 | 3.30 | 0.455 |
| 1.106 | -1.90 | 3.70 | 0.405 |
| 1.184 | -2.05 | 4.00 | 0.375 |
| 1.345 | -1.62 | 4.70 | 0.319 |
| 1.407 | -1.14 | 5.00 | 0.300 |
| 1.589 | 1.15 | 6.00 | 0.250 |
| 1.743 | 2.66 | 7.00 | 0.214 |
| 1.877 | 1.73 | 8.00 | 0.188 |
| 1.994 | 1.03 | 9.00 | 0.167 |

#### s = 9

| u | $Ne_0'(s,u)$ | y | x |
|---|---|---|---|
| 0.0000 | 1.53334 | | |
| 0.064 | 1.52 | 1.60 | 1.41 |
| 0.182 | 1.39 | 1.80 | 1.25 |
| 0.470 | 0.49 | 2.40 | 0.94 |
| 0.550 | 0.09 | 2.60 | 0.87 |
| 0.693 | -0.73 | 3.00 | 0.75 |
| 0.903 | -1.84 | 3.70 | 0.61 |
| 0.981 | -2.07 | 4.00 | 0.56 |
| 1.099 | -1.98 | 4.50 | 0.500 |
| 1.204 | -1.33 | 5.00 | 0.450 |
| 1.386 | 0.98 | 6.00 | 0.375 |
| 1.540 | 2.65 | 7.00 | 0.321 |
| 1.675 | 1.86 | 8.00 | 0.281 |
| 1.792 | -0.89 | 9.00 | 0.250 |
| 1.897 | -3.07 | 10.00 | 0.225 |
| 1.992 | -2.46 | 11.00 | 0.205 |

## $Ne'_0(s,u)$ (cont'd)

### $s = 15$

| u | $Ne'_0(s,u)$ | y | x |
|---|---|---|---|
| 0.0000 | 1.80546 | | |
| 0.032 | 1.80 | 2.00 | 1.88 |
| 0.255 | 1.16 | 2.50 | 1.50 |
| 0.438 | -0.01 | 3.00 | 1.25 |
| 0.592 | -1.18 | 3.50 | 1.07 |
| 0.725 | -1.99 | 4.00 | 0.94 |
| 0.821 | -2.21 | 4.40 | 0.85 |
| 0.949 | -1.74 | 5.00 | 0.75 |
| 1.131 | 0.56 | 6.00 | 0.62 |
| 1.285 | 2.58 | 7.00 | 0.54 |
| 1.419 | 2.15 | 8.00 | 0.469 |
| 1.536 | -0.54 | 9.00 | 0.417 |
| 1.642 | -2.98 | 10.00 | 0.375 |
| 1.737 | -2.68 | 11.00 | 0.341 |
| 1.824 | 0.30 | 12.00 | 0.312 |

## $Ne'_1(s,u)$

$$y = \frac{\sqrt{s}}{2}\, e^u, \quad x = \frac{\sqrt{s}}{2}\, e^{-u}$$

### $s = 1$

| u | $Ne'_1(s,u)$ | y | x |
|---|---|---|---|
| 0.0000 | 1.75282 | | |
| 0.182 | 1.515 | 0.60 | .417 |
| 0.470 | 1.260 | 0.80 | .312 |
| 0.588 | 1.202 | 0.90 | .278 |
| 0.693 | 1.172 | 1.00 | .250 |
| 0.876 | 1.174 | 1.20 | .208 |
| 0.956 | 1.193 | 1.30 | .192 |
| 1.224 | 1.318 | 1.70 | .147 |
| 1.435 | 1.397 | 2.10 | .119 |
| 1.609 | 1.324 | 2.50 | .100 |
| 1.723 | 1.135 | 2.80 | .089 |
| 1.792 | 0.943 | 3.00 | .083 |
| 1.974 | 0.078 | 3.60 | .069 |

### $s = 3$

| u | $Ne'_1(s,u)$ | y | x |
|---|---|---|---|
| 0.0000 | 1.22073 | | |
| 0.144 | 1.199 | 1.00 | .75 |
| 0.480 | 1.229 | 1.40 | .54 |
| 0.614 | 1.272 | 1.60 | .469 |
| 0.732 | 1.311 | 1.80 | .417 |
| 1.019 | 1.271 | 2.40 | .313 |
| 1.242 | 0.821 | 3.00 | .250 |
| 1.530 | -0.745 | 4.00 | .188 |
| 1.753 | -2.149 | 5.00 | .150 |
| 1.936 | -1.855 | 6.00 | .125 |

$$Ne'_1 (s,u) \text{ (cont'd)}$$

| | s = 6 | | | | | s = 9 | | |
|---|---|---|---|---|---|---|---|---|
| u | $Ne'_1(s,u)$ | y | x | | u | $Ne'_1(s,u)$ | y | x |
| 0.0000 | 1.13668 | | | | 0.0000 | 1.20113 | | |
| 0.060 | 1.16 | 1.30 | 1.154 | | 0.064 | 1.233 | 1.60 | 1.41 |
| 0.203 | 1.22 | 1.50 | 1.00 | | 0.182 | 1.257 | 1.80 | 1.25 |
| 0.490 | 1.27 | 2.00 | 0.75 | | 0.470 | 1.155 | 2.40 | 0.94 |
| 0.714 | 1.12 | 2.50 | 0.60 | | 0.550 | 1.033 | 2.60 | 0.87 |
| 0.827 | 0.90 | 2.80 | 0.54 | | 0.693 | 0.648 | 3.00 | 0.75 |
| 0.896 | 0.69 | 3.00 | 0.500 | | 0.903 | -0.404 | 3.70 | 0.61 |
| 0.991 | 0.30 | 3.30 | 0.455 | | 0.981 | -0.898 | 4.00 | 0.56 |
| 1.106 | -0.34 | 3.70 | 0.405 | | 1.099 | -1.675 | 4.50 | 0.500 |
| 1.184 | -0.85 | 4.00 | 0.375 | | 1.204 | -2.172 | 5.00 | 0.450 |
| 1.407 | -2.17 | 5.00 | 0.300 | | 1.386 | -1.738 | 6.00 | 0.375 |
| 1.589 | -1.78 | 6.00 | 0.250 | | 1.540 | 0.420 | 7.00 | 0.321 |
| 1.743 | 0.37 | 7.00 | 0.214 | | 1.675 | 2.545 | 8.00 | 0.281 |
| 1.877 | 2.53 | 8.00 | 0.188 | | 1.792 | 2.562 | 9.00 | 0.250 |
| 1.994 | 2.39 | 9.00 | 0.167 | | 1.897 | 0.162 | 10.00 | 0.225 |
| | | | | | 1.992 | -2.548 | 11.00 | 0.205 |

| | s = 15 | | |
|---|---|---|---|
| u | $Ne'_1(s,u)$ | y | x |
| 0.0000 | 1.43163 | | |
| 0.032 | 1.442 | 2.0 | 1.88 |
| 0.081 | 1.443 | 2.10 | 1.79 |
| 0.128 | 1.426 | 2.20 | 1.705 |
| 0.255 | 1.289 | 2.50 | 1.50 |
| 0.438 | 0.796 | 3.0 | 1.25 |
| 0.592 | 0.042 | 3.50 | 1.07 |
| 0.725 | -0.839 | 4.0 | 0.94 |
| 0.949 | -2.171 | 5.0 | 0.75 |
| 1.131 | -1.765 | 6.0 | 0.62 |
| 1.285 | 0.398 | 7.0 | 0.54 |
| 1.419 | 2.539 | 8.0 | 0.469 |
| 1.536 | 2.570 | 9.0 | 0.417 |
| 1.642 | 0.171 | 10.0 | 0.375 |
| 1.737 | -2.654 | 11.0 | 0.341 |
| 1.824 | -3.247 | 12.0 | 0.312 |

$$Ne'_2(s,u)$$

$$y = \frac{\sqrt{s}}{2} e^u, \quad x = \frac{\sqrt{s}}{2} e^{-u}$$

### s = 1

| u | $Ne'_2(s,u)$ | y | x |
|---|---|---|---|
| 0.0000 | 12.32111 | | |
| 0.182 | 8.72 | 0.60 | .417 |
| 0.470 | 4.85 | 0.80 | .312 |
| 0.588 | 3.91 | 0.90 | .278 |
| 0.693 | 3.15 | 1.00 | .250 |
| 0.876 | 2.26 | 1.20 | .208 |
| 0.956 | 1.91 | 1.30 | .192 |
| 1.224 | 1.41 | 1.70 | .147 |
| 1.435 | 1.30 | 2.10 | .119 |
| 1.609 | 1.42 | 2.50 | .100 |
| 1.723 | 1.55 | 2.80 | .089 |
| 1.792 | 1.61 | 3.00 | .083 |
| 1.974 | 1.66 | 3.60 | .069 |

### s = 3

| u | $Ne'_2(s,u)$ | y | x |
|---|---|---|---|
| 0.0000 | 3.98331 | | |
| 0.144 | 3.09 | 1.00 | .75 |
| 0.480 | 1.76 | 1.40 | .54 |
| 0.614 | 1.49 | 1.60 | .469 |
| 0.732 | 1.36 | 1.80 | .417 |
| 0.886 | 1.32 | 2.10 | .357 |
| 1.019 | 1.39 | 2.40 | .313 |
| 1.242 | 1.62 | 3.00 | .250 |
| 1.452 | 1.60 | 3.70 | .203 |
| 1.530 | 1.41 | 4.00 | .188 |
| 1.753 | -0.04 | 5.00 | .150 |
| 1.936 | -1.92 | 6.00 | .125 |

### s = 6

| u | $Ne'_2(s,u)$ | y | x |
|---|---|---|---|
| 0.0000 | 2.08246 | | |
| 0.060 | 1.93 | 1.30 | 1.15 |
| 0.203 | 1.63 | 1.50 | 1.00 |
| 0.490 | 1.37 | 2.00 | 0.75 |
| 0.714 | 1.45 | 2.50 | 0.60 |
| 0.827 | 1.55 | 2.80 | 0.54 |
| 0.896 | 1.60 | 3.00 | 0.50 |
| 0.991 | 1.62 | 3.30 | 0.455 |
| 1.106 | 1.52 | 3.70 | 0.405 |
| 1.184 | 1.31 | 4.00 | 0.375 |
| 1.345 | 0.38 | 4.70 | 0.319 |
| 1.407 | -0.16 | 5.00 | 0.300 |
| 1.589 | -1.97 | 6.00 | 0.250 |
| 1.743 | -2.45 | 7.00 | 0.214 |
| 1.877 | -0.79 | 8.00 | 0.188 |
| 1.994 | 1.85 | 9.00 | 0.167 |

### s = 9

| u | $Ne'_2(s,u)$ | y | x |
|---|---|---|---|
| 0.0000 | 1.55811 | | |
| 0.064 | 1.50 | 1.60 | 1.41 |
| 0.182 | 1.42 | 1.80 | 1.25 |
| 0.470 | 1.44 | 2.40 | 0.94 |
| 0.550 | 1.48 | 2.60 | 0.86 |
| 0.693 | 1.56 | 3.00 | 0.75 |
| 0.903 | 1.42 | 3.70 | 0.61 |
| 0.981 | 1.19 | 4.00 | 0.56 |
| 1.099 | 0.56 | 4.50 | 0.500 |
| 1.204 | -0.30 | 5.00 | 0.450 |
| 1.386 | -2.03 | 6.00 | 0.375 |
| 1.540 | -2.40 | 7.00 | 0.321 |
| 1.675 | -0.69 | 8.00 | 0.281 |
| 1.792 | 1.92 | 9.00 | 0.250 |
| 1.897 | 3.11 | 10.00 | 0.225 |
| 1.992 | 1.59 | 11.00 | 0.205 |

## $Ne'_2(s,u)$ (cont'd)

### s = 15

| u | $Ne'_2(s,u)$ | y | x |
|---|---|---|---|
| 0.0000 | 1.29635 | | |
| 0.032 | 1.31 | 2.00 | 1.88 |
| 0.255 | 1.35 | 2.50 | 1.50 |
| 0.438 | 1.42 | 3.00 | 1.25 |
| 0.592 | 1.32 | 3.50 | 1.07 |
| 0.725 | 0.94 | 4.00 | 0.94 |
| 0.821 | 0.45 | 4.40 | 0.85 |
| 0.949 | -0.55 | 5.00 | 0.75 |
| 1.131 | -2.13 | 6.00 | 0.625 |
| 1.285 | -2.30 | 7.00 | 0.536 |
| 1.419 | -0.47 | 8.00 | 0.469 |
| 1.536 | 2.07 | 9.00 | 0.417 |
| 1.642 | 3.08 | 10.00 | 0.375 |
| 1.737 | 1.41 | 11.00 | 0.341 |
| 1.824 | -1.72 | 12.00 | 0.312 |

## $No'_1(s,u)$

$$y = \frac{\sqrt{s}}{2} e^u, \quad x = \frac{\sqrt{s}}{2} e^{-u}$$

### s = 1

| u | $No'_1(s,u)$ | y | x |
|---|---|---|---|
| 0.0000 | 1.03392 | | |
| 0.182 | 0.96 | 0.60 | .417 |
| 0.470 | 0.92 | 0.80 | .312 |
| 0.588 | 0.93 | 0.90 | .278 |
| 0.693 | 0.95 | 1.00 | .250 |
| 0.876 | 1.04 | 1.20 | .208 |
| 0.956 | 1.09 | 1.30 | .192 |
| 1.224 | 1.31 | 1.70 | .147 |
| 1.435 | 1.46 | 2.10 | .119 |
| 1.609 | 1.43 | 2.50 | .100 |
| 1.723 | 1.27 | 2.80 | .089 |
| 1.792 | 1.08 | 3.00 | .083 |
| 1.974 | 0.22 | 3.60 | .069 |

### s = 3

| u | $No'_1(s,u)$ | y | x |
|---|---|---|---|
| 0.0000 | 0.32509 | | |
| 0.144 | 0.51 | 1.00 | .75 |
| 0.480 | 0.97 | 1.40 | .54 |
| 0.614 | 1.16 | 1.60 | .469 |
| 0.732 | 1.32 | 1.80 | .417 |
| 1.019 | 1.55 | 2.40 | .313 |
| 1.242 | 1.24 | 3.00 | .250 |
| 1.530 | -0.37 | 4.00 | .188 |
| 1.753 | -2.07 | 5.00 | .150 |
| 1.936 | -2.05 | 6.00 | .125 |

$$No'_1 (s, u) \quad (cont'd)$$

s = 6

| u | No'$_1$(s, u) | y | x |
|---|---|---|---|
| 0.0000 | 0.10714 | | |
| 0.060 | 0.28 | 1.30 | 1.154 |
| 0.203 | 0.67 | 1.50 | 1.00 |
| 0.490 | 1.38 | 2.00 | 0.75 |
| 0.714 | 1.66 | 2.50 | 0.60 |
| 0.827 | 1.60 | 2.80 | 0.54 |
| 0.896 | 1.47 | 3.00 | 0.500 |
| 0.991 | 1.13 | 3.30 | 0.455 |
| 1.106 | 0.47 | 3.70 | 0.405 |
| 1.184 | -0.12 | 4.00 | 0.375 |
| 1.407 | -1.99 | 5.00 | 0.300 |
| 1.589 | -2.17 | 6.00 | 0.250 |
| 1.743 | -0.18 | 7.00 | 0.214 |
| 1.877 | 2.28 | 8.00 | 0.188 |
| 1.994 | 2.82 | 9.00 | 0.167 |

s = 9

| u | No'$_1$(s, u) | y | x |
|---|---|---|---|
| 0.0000 | 0.04443 | | |
| 0.064 | 0.301 | 1.60 | 1.41 |
| 0.182 | 0.765 | 1.80 | 1.25 |
| 0.470 | 1.639 | 2.40 | 0.94 |
| 0.550 | 1.747 | 2.60 | 0.87 |
| 0.693 | 1.680 | 3.00 | 0.75 |
| 0.903 | 0.757 | 3.70 | 0.61 |
| 0.981 | 0.152 | 4.00 | 0.56 |
| 1.099 | -0.942 | 4.50 | 0.500 |
| 1.204 | -1.878 | 5.00 | 0.450 |
| 1.386 | -2.284 | 6.00 | 0.375 |
| 1.540 | -0.631 | 7.00 | 0.321 |
| 1.675 | 2.178 | 8.00 | 0.281 |
| 1.792 | 2.884 | 9.00 | 0.250 |
| 1.897 | 0.811 | 10.00 | 0.225 |
| 1.992 | -2.379 | 11.00 | 0.205 |

s = 15

| u | No'$_1$(s, u) | y | x |
|---|---|---|---|
| 0.0000 | 0.01037 | | |
| 0.032 | 0.220 | 2.00 | 1.88 |
| 0.081 | 0.518 | 2.10 | 1.79 |
| 0.128 | 0.795 | 2.20 | 1.70 |
| 0.255 | 1.461 | 2.50 | 1.50 |
| 0.438 | 1.943 | 3.00 | 1.25 |
| 0.592 | 1.658 | 3.50 | 1.07 |
| 0.725 | 0.780 | 4.00 | 0.94 |
| 0.949 | -1.561 | 5.00 | 0.75 |
| 1.131 | -2.463 | 6.00 | 0.625 |
| 1.285 | -0.812 | 7.00 | 0.54 |
| 1.419 | 1.945 | 8.00 | 0.469 |
| 1.536 | 2.994 | 9.00 | 0.417 |
| 1.642 | 1.159 | 10.00 | 0.375 |
| 1.737 | -2.013 | 11.00 | 0.341 |
| 1.824 | -3.479 | 12.00 | 0.312 |

$$No'_2(s,u)$$

$$y = \frac{\sqrt{s}}{2} e^u, \quad x = \frac{\sqrt{s}}{2} e^{-u}$$

s = 1

| u | $No'_2(s,u)$ | y | x |
|---|---|---|---|
| 0.0000 | 12.09531 | | |
| 0.182 | 8.49 | 0.60 | .417 |
| 0.470 | 4.82 | 0.80 | .312 |
| 0.588 | 3.82 | 0.90 | .278 |
| 0.693 | 3.11 | 1.00 | .250 |
| 0.876 | 2.21 | 1.20 | .208 |
| 0.956 | 1.92 | 1.30 | .192 |
| 1.224 | 1.36 | 1.70 | .147 |
| 1.435 | 1.28 | 2.10 | .119 |
| 1.609 | 1.41 | 2.50 | .100 |
| 1.723 | 1.55 | 2.80 | .089 |
| 1.792 | 1.62 | 3.00 | .083 |
| 1.974 | 1.67 | 3.60 | .069 |

s = 3

| u | $No'_2(s,u)$ | y | x |
|---|---|---|---|
| 0.0000 | 3.48552 | | |
| 0.144 | 2.73 | 1.00 | .75 |
| 0.480 | 1.59 | 1.40 | .54 |
| 0.614 | 1.36 | 1.60 | .469 |
| 0.732 | 1.26 | 1.80 | .417 |
| 1.019 | 1.36 | 2.40 | .313 |
| 1.242 | 1.62 | 3.00 | .250 |
| 1.530 | 1.47 | 4.00 | .188 |
| 1.753 | 0.02 | 5.00 | .150 |
| 1.936 | -1.89 | 6.00 | .125 |

s = 6

| u | $No'_2(s,u)$ | y | x |
|---|---|---|---|
| 0.0000 | 1.33545 | | |
| 0.060 | 1.28 | 1.30 | 1.154 |
| 0.203 | 1.16 | 1.50 | 1.00 |
| 0.490 | 1.14 | 2.00 | 0.75 |
| 0.714 | 1.37 | 2.50 | 0.60 |
| 0.827 | 1.53 | 2.80 | 0.54 |
| 0.896 | 1.63 | 3.00 | 0.500 |
| 0.991 | 1.71 | 3.30 | 0.455 |
| 1.106 | 1.67 | 3.70 | 0.405 |
| 1.184 | 1.50 | 4.00 | 0.375 |
| 1.407 | 0.06 | 5.00 | 0.300 |
| 1.589 | -1.87 | 6.00 | 0.250 |
| 1.743 | -2.52 | 7.00 | 0.214 |
| 1.877 | -0.96 | 8.00 | 0.188 |
| 1.994 | 1.72 | 9.00 | 0.167 |

s = 9

| u | $No'_2(s,u)$ | y | x |
|---|---|---|---|
| 0.0000 | 0.66117 | | |
| 0.064 | 0.71 | 1.60 | 1.41 |
| 0.182 | 0.83 | 1.80 | 1.25 |
| 0.470 | 1.23 | 2.40 | 0.94 |
| 0.550 | 1.38 | 2.60 | 0.87 |
| 0.693 | 1.62 | 3.00 | 0.75 |
| 0.903 | 1.72 | 3.70 | 0.61 |
| 0.981 | 1.56 | 4.00 | 0.56 |
| 1.099 | 0.99 | 4.50 | 0.500 |
| 1.204 | 0.12 | 5.00 | 0.450 |
| 1.386 | -1.84 | 6.00 | 0.375 |
| 1.540 | -2.53 | 7.00 | 0.321 |
| 1.675 | -1.00 | 8.00 | 0.281 |
| 1.792 | 1.69 | 9.00 | 0.250 |
| 1.897 | 3.14 | 10.00 | 0.225 |
| 1.992 | 1.81 | 11.00 | 0.205 |

$$No'_2(s,u) \quad (cont'd)$$

$$s = 15$$

| u | $No'_2(s,u)$ | y | x |
|---|---|---|---|
| 0.0000 | 0.20909 | | |
| 0.032 | 0.32 | 2.00 | 1.875 |
| 0.081 | 0.48 | 2.10 | 1.785 |
| 0.128 | 0.63 | 2.20 | 1.705 |
| 0.255 | 1.04 | 2.50 | 1.50 |
| 0.438 | 1.57 | 3.00 | 1.25 |
| 0.592 | 1.83 | 3.50 | 1.07 |
| 0.725 | 1.72 | 4.00 | 0.94 |
| 0.949 | 0.31 | 5.00 | 0.75 |
| 1.131 | −1.74 | 6.00 | 0.625 |
| 1.285 | −2.58 | 7.00 | 0.54 |
| 1.419 | −1.12 | 8.00 | 0.469 |
| 1.536 | 1.58 | 9.00 | 0.417 |
| 1.642 | 3.14 | 10.00 | 0.375 |
| 1.737 | 1.90 | 11.00 | 0.341 |
| 1.824 | −1.25 | 12.00 | 0.312 |

# TABLE IV

Derivatives with Respect to u of the
Radial Mathieu Functions of the Third Kind

$$He_o^{(1)'}(-|s|,u)$$

$$y = \frac{\sqrt{|s|}}{2}\, e^u, \quad x = \frac{\sqrt{|s|}}{2}\, e^{-u}$$

### s = 1

| u | $He_o^{(1)'}(-|s|,u)$ | y | x | u | $He_o^{(1)'}(-|s|,u)$ | y | x |
|---|---|---|---|---|---|---|---|
| 0.000 | .700 36i | | | 0.956 | .383i | 1.30 | .19 |
| 0.182 | .642i | 0.60 | .42 | 1.224 | .282i | 1.70 | .15 |
| 0.470 | .551i | 0.80 | .31 | 1.435 | .204i | 2.10 | .12 |
| 0.693 | .478i | 1.00 | .25 | 1.723 | .114i | 2.80 | .09 |
| | | | | 2.197 | .025i | 4.50 | .01 |

### s = 3

| u | $He_o^{(1)'}(-|s|,u)$ | y | x |
|---|---|---|---|
| 0.000 | .52782i | | |
| 0.326 | .393i | 1.20 | .62 |
| 0.480 | .336i | 1.40 | .54 |
| 0.614 | .288i | 1.60 | .47 |
| 0.732 | .246i | 1.80 | .42 |
| 0.144 | .464i | 1.00 | .75 |
| 0.837 | .210i | 2.00 | .38 |
| 1.019 | .152i | 2.40 | .31 |
| 1.099 | .129i | 2.60 | .29 |
| 1.466 | .048i | 3.75 | .20 |

### s = 6

| u | $He_o^{(1)'}(-|s|,u)$ | y | x |
|---|---|---|---|
| 0.000 | .34210i | | |
| 0.060 | .318i | 1.30 | 1.15 |
| 0.203 | .268i | 1.50 | 1.00 |
| 0.490 | .182i | 2.00 | 0.75 |
| 0.714 | .123i | 2.50 | 0.60 |
| 0.827 | .097i | 2.80 | 0.54 |
| 0.896 | .082i | 3.00 | 0.50 |
| 0.991 | .064i | 3.30 | 0.45 |
| 1.106 | .046i | 3.70 | 0.40 |
| 1.407 | .015i | 5.00 | 0.30 |

### s = 9

| u | $He_o^{(1)'}(-|s|,u)$ | y | x |
|---|---|---|---|
| 0.000 | .22757 | | |
| 0.064 | .208i | 1.60 | 1.41 |
| 0.182 | .177i | 1.80 | 1.25 |
| 0.288 | .152i | 2.00 | 1.12 |
| 0.470 | .113i | 2.40 | 0.94 |
| 0.550 | .097i | 2.60 | 0.87 |
| 0.693 | .071i | 3.00 | 0.75 |
| 0.903 | .041i | 3.70 | 0.61 |
| 0.981 | .032i | 4.00 | 0.56 |
| 1.099 | .021i | 4.50 | 0.50 |

### s = 15

| u | $He_o^{(1)'}(-|s|,u)$ | y | x |
|---|---|---|---|
| 0.000 | .11108i | | |
| 0.032 | .1052i | 2.00 | 1.875 |
| 0.255 | .0729i | 2.50 | 1.50 |
| 0.438 | .0514i | 3.00 | 1.25 |
| 0.725 | .0243i | 4.00 | 0.94 |

$$He_1^{(1)'}(-|s|,u)$$
$$y = \frac{\sqrt{|s|}}{2} e^u, \quad x = \frac{\sqrt{|s|}}{2} e^{-u}$$

### s = 1

| u | $He_1^{(1)'}(-|s|,u)$ | y | x | u | $He_1^{(1)'}(-|s|,u)$ | y | x |
|---|---|---|---|---|---|---|---|
| 0.000 | 1.45326 | | | 0.956 | 0.531 | 1.30 | .19 |
| 0.182 | 1.214 | 0.60 | .42 | 1.224 | 0.361 | 1.70 | .15 |
| 0.470 | 0.916 | 0.80 | .31 | 1.435 | 0.250 | 2.10 | .12 |
| 0.693 | 0.724 | 1.00 | .25 | 1.723 | 0.132 | 2.80 | .09 |
| | | | | 2.197 | 0.028 | 4.50 | .01 |

### s = 3

| u | $He_1^{(1)'}(-|s|,u)$ | y | x |
|---|---|---|---|
| 0.000 | .69715 | | |
| 0.144 | .594 | 1.00 | .75 |
| 0.326 | .484 | 1.20 | .62 |
| 0.480 | .402 | 1.40 | .54 |
| 0.614 | .337 | 1.60 | .47 |
| 0.732 | .283 | 1.80 | .42 |
| 0.837 | .238 | 2.00 | .38 |
| 1.019 | .169 | 2.40 | .31 |
| 1.099 | .142 | 2.60 | .29 |
| 1.466 | .051 | 3.75 | .20 |

### s = 6

| u | $He_1^{(1)'}(-|s|,u)$ | y | x |
|---|---|---|---|
| 0.000 | .37590 | | |
| 0.060 | .348 | 1.30 | 1.154 |
| 0.203 | .290 | 1.50 | 1.00 |
| 0.490 | .193 | 2.00 | 0.75 |
| 0.714 | .129 | 2.50 | 0.60 |
| 0.827 | .101 | 2.80 | 0.54 |
| 0.896 | .085 | 3.00 | 0.50 |
| 0.991 | .066 | 3.30 | 0.45 |
| 1.106 | .047 | 3.70 | 0.40 |
| 1.407 | .015 | 5.00 | 0.30 |

### s = 9

| u | $He_1^{(1)'}(-|s|,u)$ | y | x |
|---|---|---|---|
| 0.000 | .23637 | | |
| 0.064 | .216 | 1.60 | 1.41 |
| 0.182 | .182 | 1.80 | 1.25 |
| 0.288 | .156 | 2.00 | 1.12 |
| 0.470 | .115 | 2.40 | 0.94 |
| 0.550 | .099 | 2.60 | 0.87 |
| 0.693 | .073 | 3.00 | 0.75 |
| 0.903 | .041 | 3.70 | 0.61 |
| 0.981 | .032 | 4.00 | 0.56 |
| 1.099 | .021 | 4.50 | 0.50 |

### s = 15

| u | $He_1^{(1)'}(-|s|,u)$ | y | x |
|---|---|---|---|
| 0.000 | .11200 | | |
| 0.032 | .1057 | 2.0 | 1.875 |
| 0.255 | .0734 | 2.5 | 1.50 |
| 0.438 | .0516 | 3.0 | 1.25 |
| 0.725 | .0243 | 4.0 | 0.94 |

$$He_2^{(1)'}(-|s|, u)$$

$$y = \frac{\sqrt{|s|}}{2} e^u, \quad x = \frac{\sqrt{|s|}}{2} e^{-u}$$

### s = 1

| u | $He_2^{(1)'}(-|s|,u)$ | y | x | u | $He_2^{(1)'}(-|s|,u)$ | y | x |
|---|---|---|---|---|---|---|---|
| 0.000 | -13.38025i | | | 0.956 | -1.8i | 1.30 | .19 |
| 0.182 | -9.2i | 0.60 | .42 | 1.224 | -1.0i | 1.70 | .15 |
| 0.470 | -5.0i | 0.80 | .31 | 1.435 | -0.6i | 2.10 | .12 |
| 0.693 | -3.1i | 1.00 | .25 | 1.723 | -0.2i | 2.80 | .09 |
| | | | | 2.197 | -0.0i | 4.50 | .01 |

### s = 3

| u | $He_2^{(1)'}(-|s|,u)$ | y | x |
|---|---|---|---|
| 0.000 | -5.00660i | | |
| 0.144 | -3.58i | 1.00 | .75 |
| 0.326 | -2.35i | 1.20 | .62 |
| 0.480 | -1.64i | 1.40 | .54 |
| 0.614 | -1.18i | 1.60 | .47 |
| 0.732 | -0.89i | 1.80 | .42 |
| 0.837 | -0.67i | 2.00 | .38 |
| 1.019 | -0.40i | 2.40 | .31 |
| 1.099 | -0.32i | 2.60 | .29 |
| 1.466 | -0.09i | 3.75 | .20 |

### s = 6

| u | $He_2^{(1)'}(-|s|,u)$ | y | x |
|---|---|---|---|
| 0.000 | -2.99995i | | |
| 0.060 | -2.55i | 1.30 | 1.15 |
| 0.203 | -1.73i | 1.50 | 1.00 |
| 0.490 | -0.79i | 2.00 | 0.75 |
| 0.714 | -0.41i | 2.50 | 0.60 |
| 0.827 | -0.28i | 2.80 | 0.54 |
| 0.896 | -0.22i | 3.00 | 0.50 |
| 0.991 | -0.16i | 3.30 | 0.45 |
| 1.106 | -0.10i | 3.70 | 0.40 |
| 1.407 | -0.03i | 5.00 | 0.30 |

### s = 9

| u | $He_2^{(1)'}(-|s|,u)$ | y | x |
|---|---|---|---|
| 0.000 | -2.32737i | | |
| 0.064 | -1.92i | 1.60 | 1.41 |
| 0.182 | -1.33i | 1.80 | 1.25 |
| 0.288 | -0.96i | 2.00 | 1.12 |
| 0.470 | -0.54i | 2.40 | 0.94 |
| 0.550 | -0.42i | 2.60 | 0.87 |
| 0.693 | -0.25i | 3.00 | 0.75 |
| 0.903 | -0.11i | 3.70 | 0.61 |
| 0.981 | -0.08i | 4.00 | 0.56 |
| 1.099 | -0.05i | 4.50 | 0.50 |

### s = 15

| u | $He_2^{(1)'}(-|s|,u)$ | y | x |
|---|---|---|---|
| 0.000 | -1.68637i | | |
| 0.032 | -1.5004i | 2.00 | 1.875 |
| 0.255 | -0.6581i | 2.50 | 1.50 |
| 0.438 | -0.3290i | 3.00 | 1.25 |
| 0.725 | -0.0986i | 4.00 | 0.94 |

$$\cdot Ho_1^{(1)'}(-|s|, u)$$

$$y = \frac{\sqrt{|s|}}{2} e^u, \quad x = \frac{\sqrt{|s|}}{2} e^{-u}$$

s = 1

| u | $Ho_1^{(1)'}(-|s|,u)$ | y | x | u | $Ho_1^{(1)'}(-|s|,u)$ | y | x |
|---|---|---|---|---|---|---|---|
| 0.000 | 2.180 | 0.50 | .50 | 0.956 | 0.65 | 1.30 | .19 |
| 0.182 | 1.75 | 0.60 | .42 | 1.224 | 0.42 | 1.70 | .15 |
| 0.470 | 1.23 | 0.80 | .31 | 1.435 | 0.28 | 2.10 | .12 |
| 0.693 | 0.93 | 1.00 | .25 | 1.723 | 0.15 | 2.80 | .09 |
|  |  |  |  | 2.197 | 0.03 | 4.50 | .01 |

s = 3 ; s = 6

| u | $Ho_1^{(1)'}(-|s|,u)$ | y | x | u | $Ho_1^{(1)'}(-|s|,u)$ | y | x |
|---|---|---|---|---|---|---|---|
| 0.000 | 1.566 | 0.866 | .866 | 0.000 | 1.226 | 1.225 | 1.225 |
| 0.144 | 1.232 | 1.00 | .75 | 0.060 | 1.074 | 1.30 | 1.154 |
| 0.326 | 0.911 | 1.20 | .62 | 0.203 | 0.793 | 1.50 | 1.00 |
| 0.480 | 0.702 | 1.40 | .54 | 0.490 | 0.419 | 2.00 | 0.75 |
| 0.614 | 0.551 | 1.60 | .47 | 0.714 | 0.241 | 2.50 | 0.60 |
| 0.732 | 0.441 | 1.80 | .42 | 0.827 | 0.177 | 2.80 | 0.54 |
| 0.837 | 0.355 | 2.00 | .38 | 0.896 | 0.144 | 3.00 | 0.50 |
| 1.019 | 0.235 | 2.40 | .31 | 0.991 | 0.107 | 3.30 | 0.45 |
| 1.099 | 0.193 | 2.60 | .29 | 1.106 | 0.072 | 3.70 | 0.40 |
| 1.466 | 0.064 | 3.75 | .20 | 1.407 | 0.020 | 5.00 | 0.30 |

s = 9 ; s = 15

| u | $Ho_1^{(1)'}(-|s|,u)$ | y | x | u | $Ho_1^{(1)'}(-|s|,u)$ | y | x |
|---|---|---|---|---|---|---|---|
| 0.000 | .987 | 1.50 | 1.50 | 0.000 | .6288 | 1.936 | 1.936 |
| 0.064 | .842 | 1.60 | 1.41 | 0.032 | .5737 | 2.00 | 1.875 |
| 0.182 | .625 | 1.80 | 1.25 | 0.255 | .2899 | 2.50 | 1.50 |
| 0.288 | .478 | 2.00 | 1.12 | 0.438 | .1625 | 3.00 | 1.25 |
| 0.470 | .296 | 2.40 | 0.94 | 0.725 | .0572 | 4.00 | 0.94 |
| 0.550 | .237 | 2.60 | 0.87 |  |  |  |  |
| 0.693 | .154 | 3.00 | 0.75 |  |  |  |  |
| 0.903 | .076 | 3.70 | 0.61 |  |  |  |  |
| 0.981 | .056 | 4.00 | 0.56 |  |  |  |  |
| 1.099 | .034 | 4.50 | 0.50 |  |  |  |  |

$$Ho_2^{(1)'}(\sim|s|,u)$$

$$y = \frac{\sqrt{|s|}}{2}\, e^u, \quad x = \frac{\sqrt{|s|}}{2}\, e^{-u}$$

s = 1

| u | $Ho_2^{(1)'}(-|s|,u)$ | y | x | u | $Ho_2^{(1)'}(-|s|,u)$ | y | x |
|---|---|---|---|---|---|---|---|
| 0.000 | -13.145i | 0.50 | .50 | 0.956 | -1.75i | 1.30 | .19 |
| 0.182 | -8.99i | 0.60 | .42 | 1.224 | -0.94i | 1.70 | .15 |
| 0.470 | -4.94i | 0.80 | .31 | 1.435 | -0.55i | 2.10 | .12 |
| 0.693 | -3.09i | 1.00 | .25 | 1.723 | -0.24i | 2.80 | .09 |
| | | | | 2.197 | -0.04i | 4.50 | .01 |

s = 3

| u | $Ho_2^{(1)'}(-|s|,u)$ | y | x |
|---|---|---|---|
| 0.000 | -4.475i | 0.866 | .866 |
| 0.144 | -3.229i | 1.00 | .75 |
| 0.326 | -2.138i | 1.20 | .62 |
| 0.480 | -1.522i | 1.40 | .54 |
| 0.614 | -1.096i | 1.60 | .47 |
| 0.732 | -0.822i | 1.80 | .42 |
| 0.837 | -0.629i | 2.00 | .38 |
| 1.019 | -0.383i | 2.40 | .31 |
| 1.099 | -0.303i | 2.60 | .29 |
| 1.466 | -0.087i | 3.75 | .20 |

s = 6

| u | $Ho_2^{(1)'}(-|s|,u)$ | y | x |
|---|---|---|---|
| 0.000 | -2.210i | 1.225 | 1.225 |
| 0.060 | -1.889i | 1.30 | 1.154 |
| 0.203 | -1.319i | 1.50 | 1.00 |
| 0.490 | -0.629i | 2.00 | 0.75 |
| 0.714 | -0.337i | 2.50 | 0.60 |
| 0.827 | -0.239i | 2.80 | 0.54 |
| 0.896 | -0.191i | 3.00 | 0.50 |
| 0.991 | -0.138i | 3.30 | 0.45 |
| 1.106 | -0.091i | 3.70 | 0.40 |
| 1.407 | -0.024i | 5.00 | 0.30 |

s = 9

| u | $Ho_2^{(1)'}(-|s|,u)$ | y | x |
|---|---|---|---|
| 0.000 | -1.393i | 1.50 | 1.50 |
| 0.064 | -1.166i | 1.60 | 1.41 |
| 0.182 | -0.844i | 1.80 | 1.25 |
| 0.288 | -0.631i | 2.00 | 1.12 |
| 0.470 | -0.376i | 2.40 | 0.94 |
| 0.550 | -0.296i | 2.60 | 0.87 |
| 0.693 | -0.187i | 3.00 | 0.75 |
| 0.903 | -0.089i | 3.70 | 0.61 |
| 0.981 | -0.065i | 4.00 | 0.56 |
| 1.099 | -0.039i | 4.50 | 0.50 |

s = 15

| u | $Ho_2^{(1)'}(-|s|,u)$ | y | x |
|---|---|---|---|
| 0.000 | -.7063i | 1.936 | 1.936 |
| 0.032 | -.6397i | 2.00 | 1.875 |
| 0.255 | -.3200i | 2.50 | 1.50 |
| 0.438 | -.1769i | 3.00 | 1.25 |
| 0.725 | -.0610i | 4.00 | 0.94 |

## TABLE V

### Periodic Mathieu Functions of Order Two
### for Negative Real Values of Parameter s.

$$Se_2(-|s|, v) = \frac{1}{Se_2(|s|, \pi/2)} \sum_{k=0}^{\infty} (-1)^k De_{2k} \cos 2kv$$

| v | $|s| = 1$ | 3 | 6 | 9 | 15 |
|---|---|---|---|---|---|
| $0^o$ | +1.0000 | +1.0000 | +1.0000 | +1.0000 | +1.0000 |
| $15^o$ | +0.8492 | +0.8114 | +0.7481 | +0.6827 | +0.5577 |
| $30^o$ | +0.4456 | +0.3254 | +0.1338 | -0.0505 | -0.3619 |
| $45^o$ | -0.0865 | -0.2721 | -0.5456 | -0.7758 | -1.0728 |
| $60^o$ | -0.5969 | -0.7979 | -1.0630 | -1.2389 | -1.3341 |
| $75^o$ | -0.9570 | -1.1402 | -1.3520 | -1.4432 | -1.3315 |
| $90^o$ | -1.0860 | -1.2569 | -1.4406 | -1.4937 | -1.3009 |

$$So_2(-|s|, v) = \frac{1}{So_2'(|s|, \pi/2)} \sum_{k=1}^{\infty} (-1)^k Do_{2k} \sin 2kv$$

| v | $|s| = 1$ | 3 | 6 | 9 | 15 |
|---|---|---|---|---|---|
| $0^o$ | 0.0000 | 0.0000 | 0.0000 | 0.0000 | 0.0000 |
| $15^o$ | 0.2486 | 0.2460 | 0.2422 | 0.2388 | 0.2327 |
| $30^o$ | 0.4242 | 0.4074 | 0.3845 | 0.3640 | 0.3292 |
| $45^o$ | 0.4797 | 0.4422 | 0.3928 | 0.3507 | 0.2840 |
| $60^o$ | 0.4069 | 0.3596 | 0.2998 | 0.2513 | 0.1798 |
| $75^o$ | 0.2313 | 0.1981 | 0.1574 | 0.1255 | 0.0811 |
| $90^o$ | 0.0000 | 0.0000 | 0.0000 | 0.0000 | 0.0000 |

## TABLE VI

Positions of the First Zeros of the Radial Mathieu Functions
of the First Kind, not Counting Zeros Occurring at u = 0.

Values of u

|  | s = 1 | 3 | 6 | 9 | 15 | 25 | 40 |
|---|---|---|---|---|---|---|---|
| $Je_0(s,u)$ | 1.573 | 1.044 | 0.748 | 0.603 | 0.457 | 0.346 | 0.268 |
| $Je_1(s,u)$ |  | 1.464 | 1.100 | 0.889 | 0.643 | 0.451 | 0.328 |
| $Je_2(s,u)$ |  | 1.776 | 1.418 | 1.202 | 0.920 | 0.640 | 0.431 |
| $Jo_1(s,u)$ |  | 1.514 | 1.196 | 1.022 | 0.822 | 0.648 | 0.516 |
| $Jo_2(s,u)$ |  | 1.781 | 1.438 | 1.240 | 1.000 | 0.780 | 0.606 |

## TABLE VII

Positions of the Second Zeros of the Radial Mathieu Functions
of the First Kind, not Counting Zeros Occurring at u=0.

Values of u

|  | s = 3 | 6 | 9 | 15 | 25 | 40 |
|---|---|---|---|---|---|---|
| $Je_0(s,u)$ | 1.856 | 1.521 | 1.332 | 1.107 | 0.901 | 0.734 |
| $Je_1(s,u)$ |  | 1.734 | 1.528 | 1.274 | 1.034 | 0.835 |
| $Je_2(s,u)$ |  | 1.922 | 1.715 | 1.450 | 1.185 | 0.954 |
| $Jo_1(s,u)$ |  | 1.763 | 1.570 | 1.334 | 1.112 | 0.924 |
| $Jo_2(s,u)$ |  | 1.929 | 1.728 | 1.478 | 1.235 | 1.025 |

## TABLE VIII

Positions of the First Zeros of the Derivatives
with Respect to u of the Radial Mathieu Functions
of the First Kind, not Counting Zeros Occurring at u = 0.

Values of u

|  | s = 1 | 3 | 6 | 9 | 15 | 25 | 40 |
|---|---|---|---|---|---|---|---|
| $Je_0'(s,u)$ |  | 1.497 | 1.175 | 1.001 | 0.806 |  |  |
| $Je_1'(s,u)$ | 1.227 | 0.409 | 1.449 | 1.241 | 0.989 | 0.776 | 0.596 |
| $Je_2'(s,u)$ | 1.807 | 1.240 | 0.834 | 0.526 | 1.209 |  |  |
| $Jo_1'(s,u)$ | 1.365 | 0.913 | 0.677 | 0.560 | 0.436 | 0.342 | 0.264 |
| $Jo_2'(s,u)$ | 1.811 | 1.268 | 0.941 | 0.766 | 0.574 |  |  |

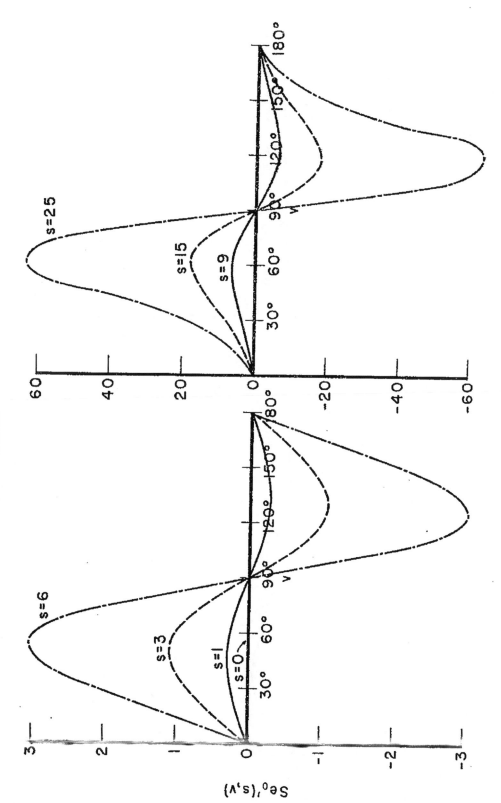

FIGURE 1    DERIVATIVE WITH RESPECT TO v OF THE EVEN PERIODIC MATHIEU
FUNCTION OF ORDER ZERO

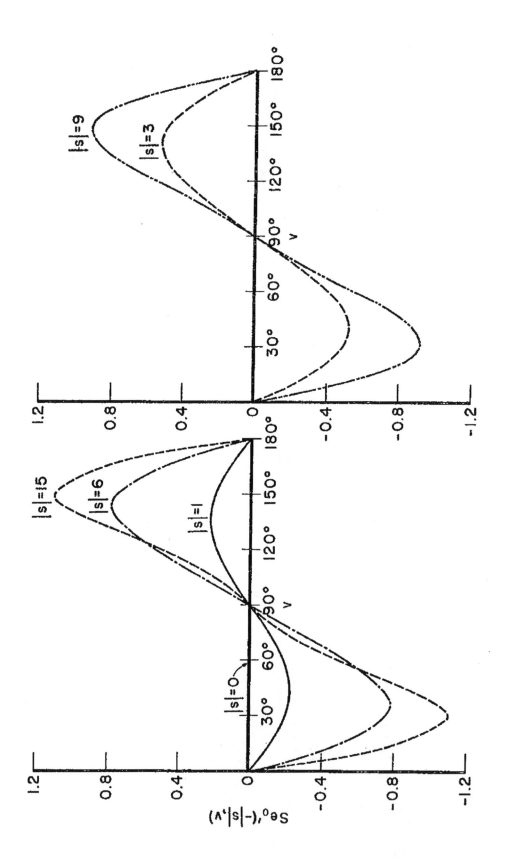

FIGURE 2    DERIVATIVE WITH RESPECT TO v OF THE EVEN PERIODIC MATHIEU
FUNCTION OF ORDER ZERO WITH NEGATIVE PARAMETER "s"

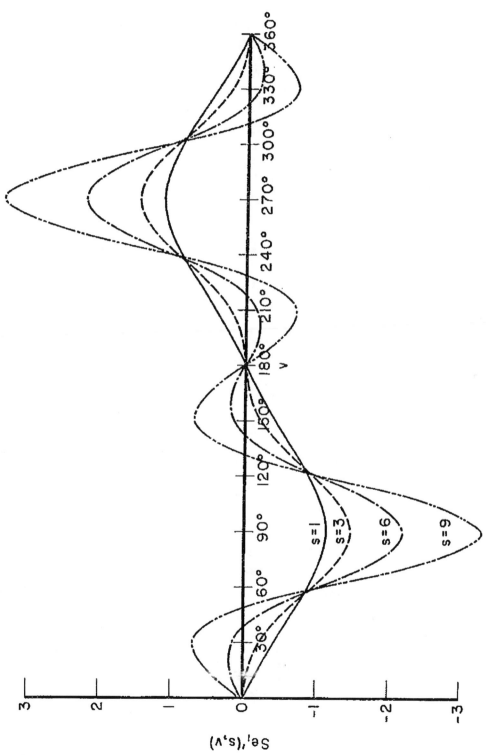

FIGURE 3    DERIVATIVE WITH RESPECT TO v OF THE EVEN PERIODIC MATHIEU
FUNCTION OF ORDER ONE

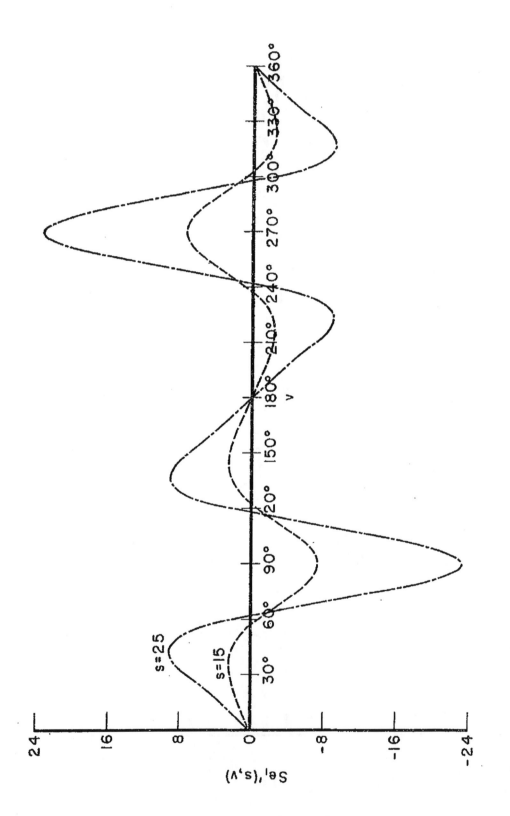

FIGURE 4    DERIVATIVE WITH RESPECT TO v OF THE EVEN PERIODIC MATHIEU
FUNCTION OF ORDER ONE

FIGURE 5    DERIVATIVE WITH RESPECT TO $v$ OF THE EVEN PERIODIC MATHIEU
FUNCTION OF ORDER ONE WITH NEGATIVE PARAMETER "s"

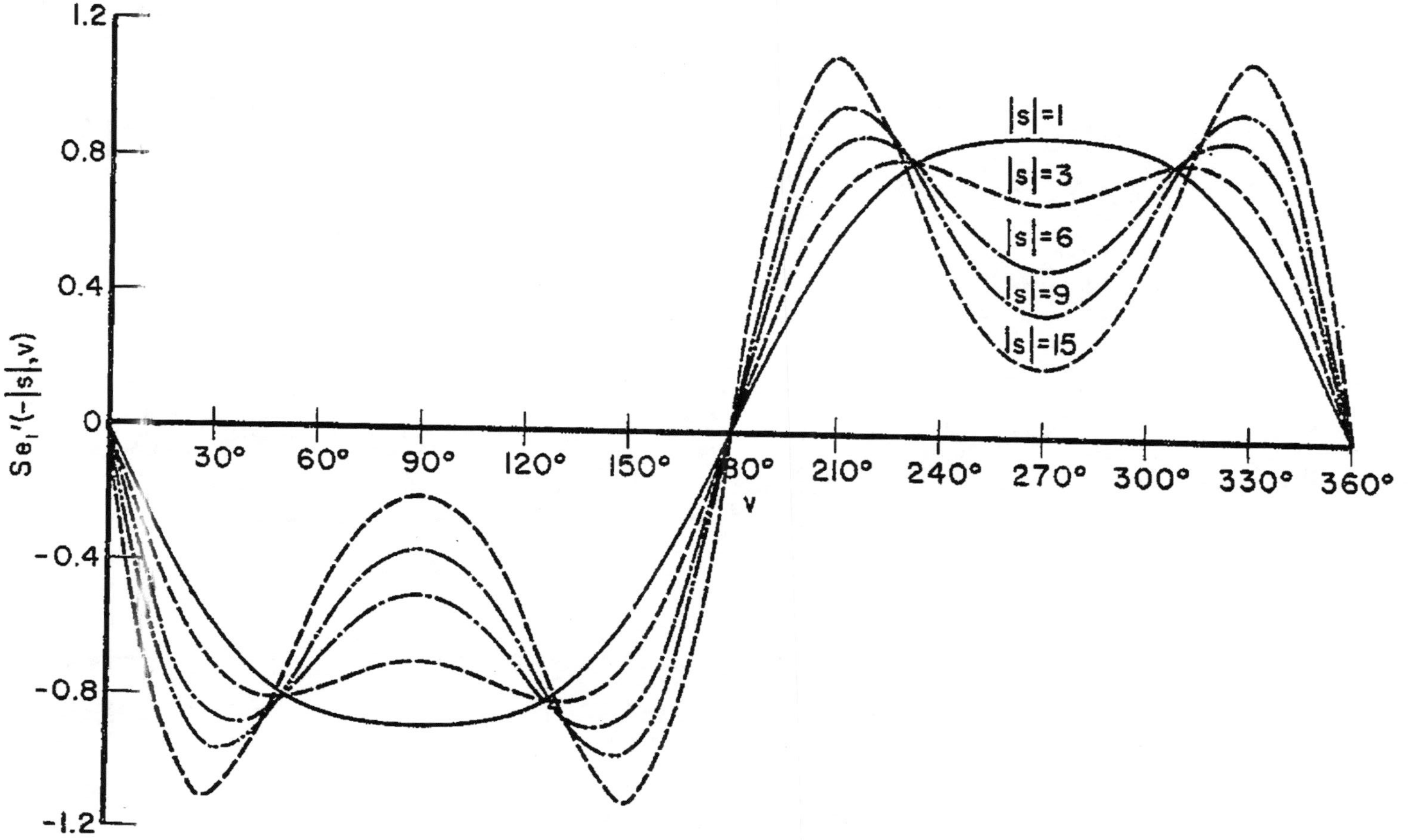

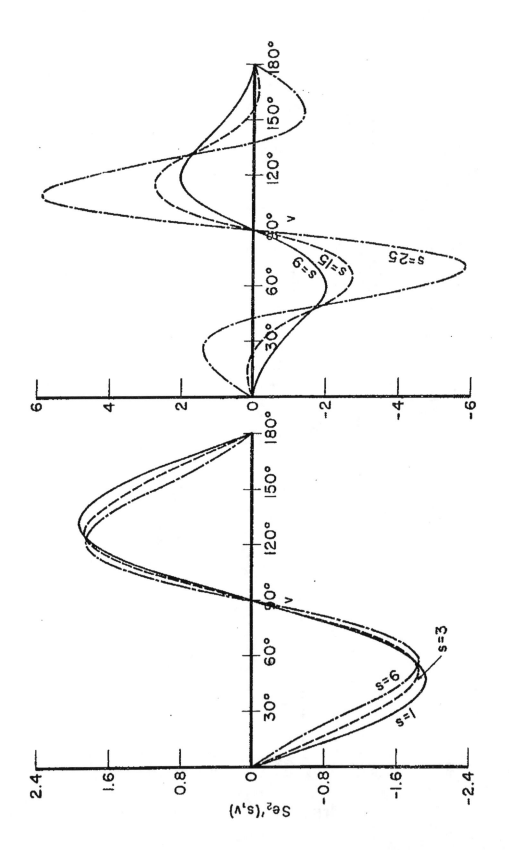

FIGURE 6    DERIVATIVE WITH RESPECT TO v OF THE EVEN PERIODIC MATHIEU
FUNCTION OF ORDER TWO

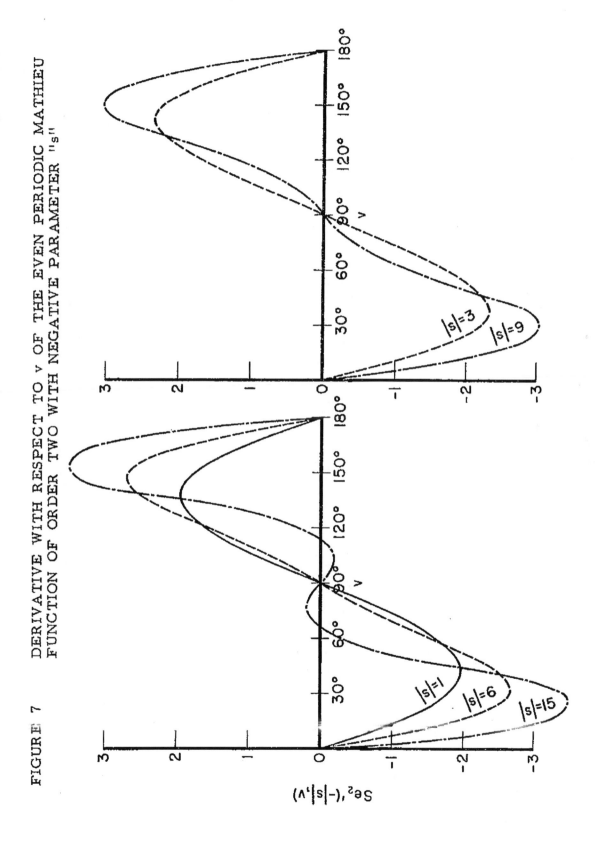

FIGURE 7    DERIVATIVE WITH RESPECT TO v OF THE EVEN PERIODIC MATHIEU
FUNCTION OF ORDER TWO WITH NEGATIVE PARAMETER "s"

141

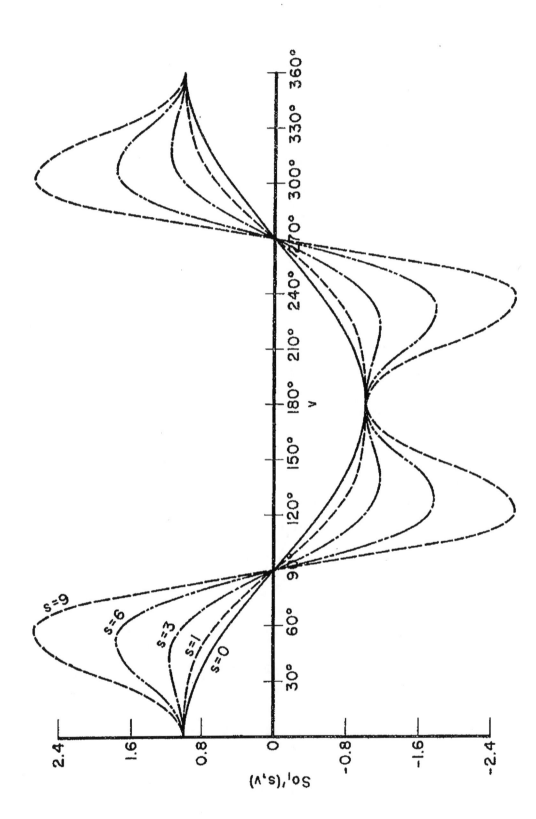

FIGURE 8    DERIVATIVE WITH RESPECT TO v OF THE ODD PERIODIC MATHIEU
FUNCTION OF ORDER ONE

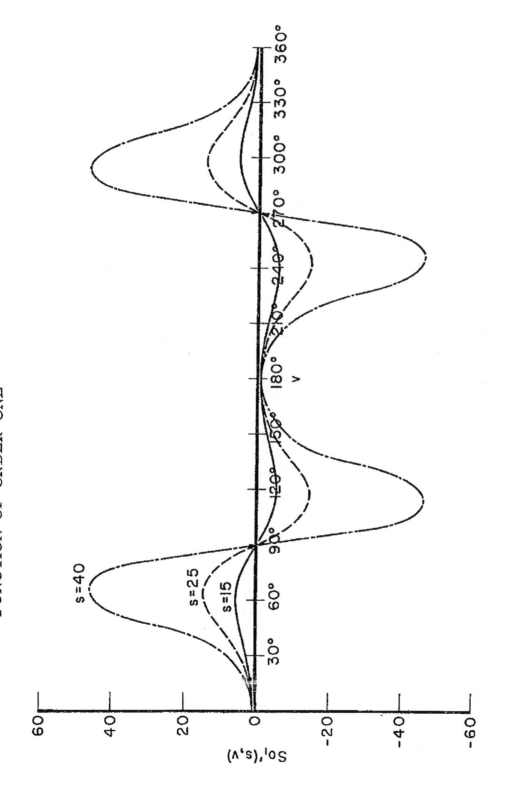

FIGURE 9    DERIVATIVE WITH RESPECT TO v OF THE ODD PERIODIC MATHIEU
FUNCTION OF ORDER ONE

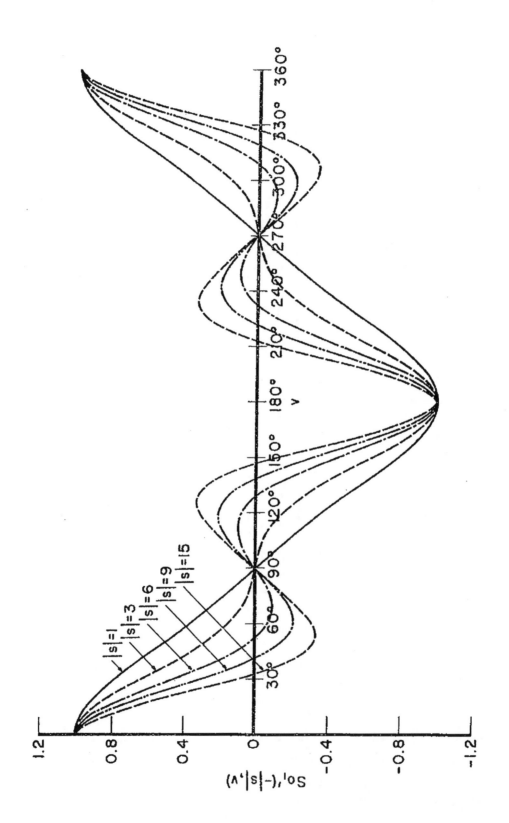

FIGURE 10    DERIVATIVE WITH RESPECT TO v OF THE ODD PERIODIC MATHIEU
FUNCTION OF ORDER ONE WITH NEGATIVE PARAMETER "s"

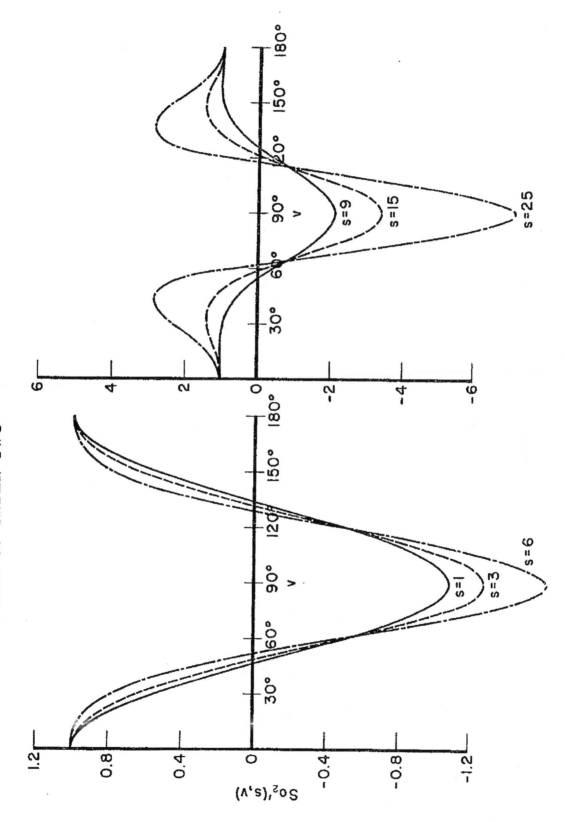

FIGURE 11  DERIVATIVE WITH RESPECT TO v OF THE ODD PERIODIC MATHIEU
FUNCTION OF ORDER TWO

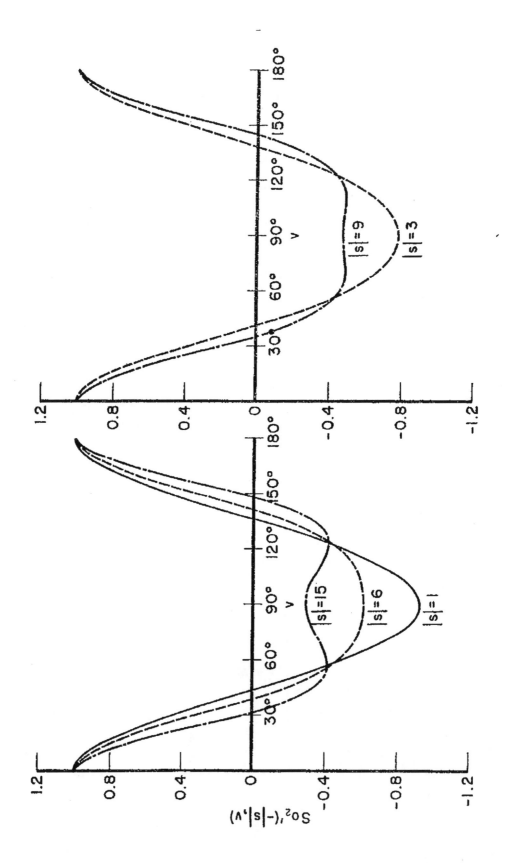

FIGURE 12    DERIVATIVE WITH RESPECT TO $v$ OF THE ODD PERIODIC MATHIEU
FUNCTION OF ORDER TWO WITH NEGATIVE PARAMETER "s"

FIGURE 13   DERIVATIVE WITH RESPECT TO u OF THE "EVEN" RADIAL MATHIEU
FUNCTION OF THE FIRST KIND AND ORDER ZERO

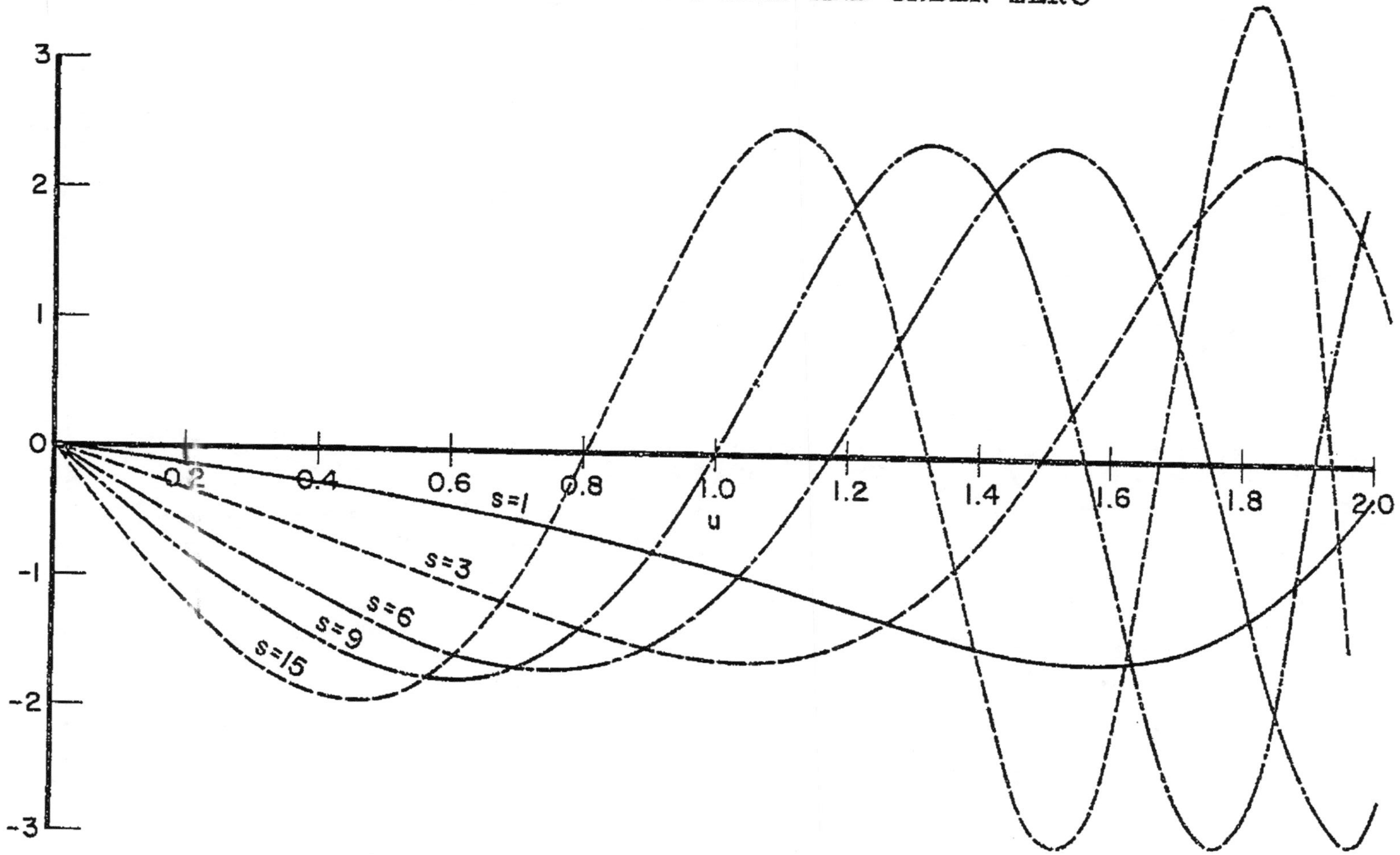

147

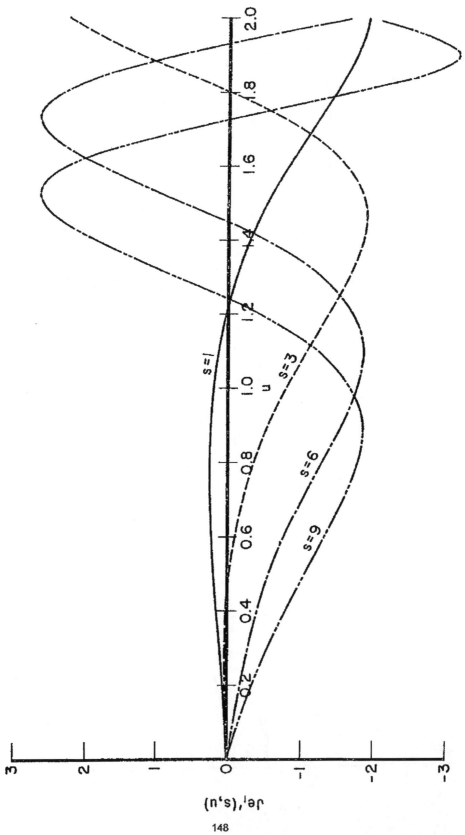

FIGURE 14   DERIVATIVE WITH RESPECT TO u OF THE "EVEN" RADIAL MATHIEU
FUNCTION OF THE FIRST KIND AND ORDER ONE

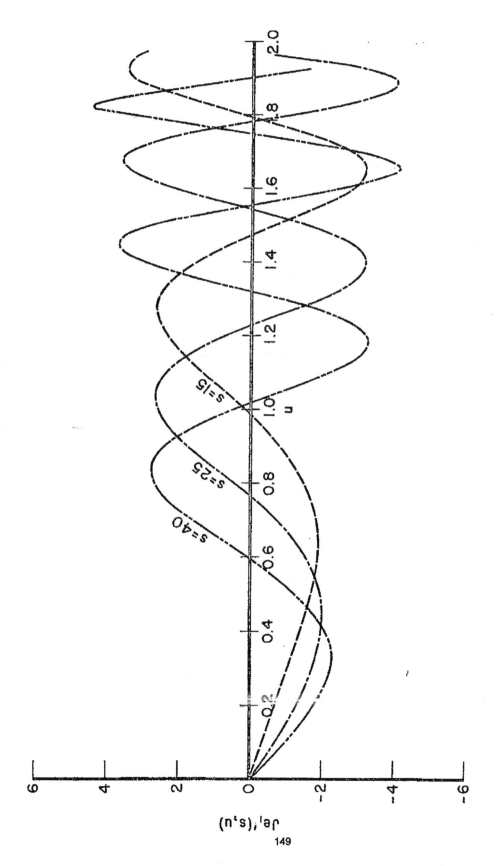

FIGURE 15    DERIVATIVE WITH RESPECT TO u OF THE "EVEN" RADIAL MATHIEU
FUNCTION OF THE FIRST KIND AND ORDER ONE

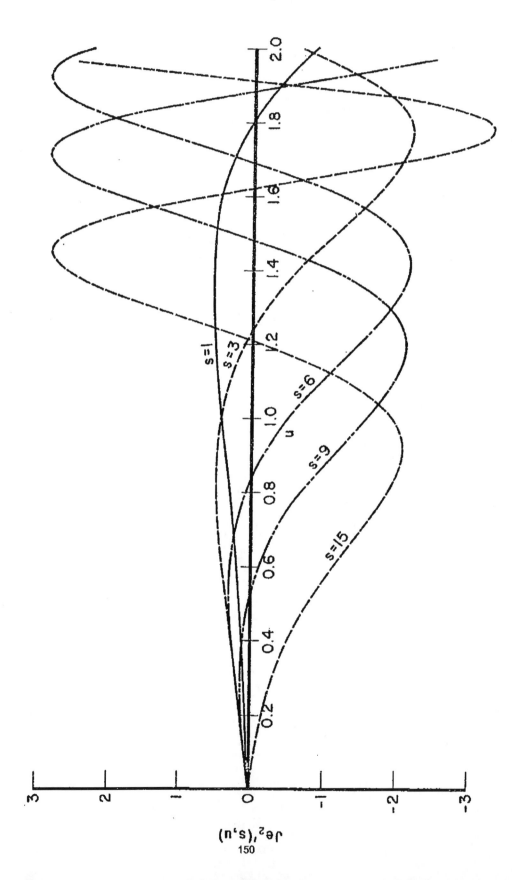

FIGURE 16    DERIVATIVE WITH RESPECT TO u OF THE "EVEN" RADIAL MATHIEU
FUNCTION OF THE FIRST KIND AND ORDER TWO

150

FIGURE 17    DERIVATIVE WITH RESPECT TO u OF THE "ODD" RADIAL MATHIEU
FUNCTION OF THE FIRST KIND AND ORDER ONE

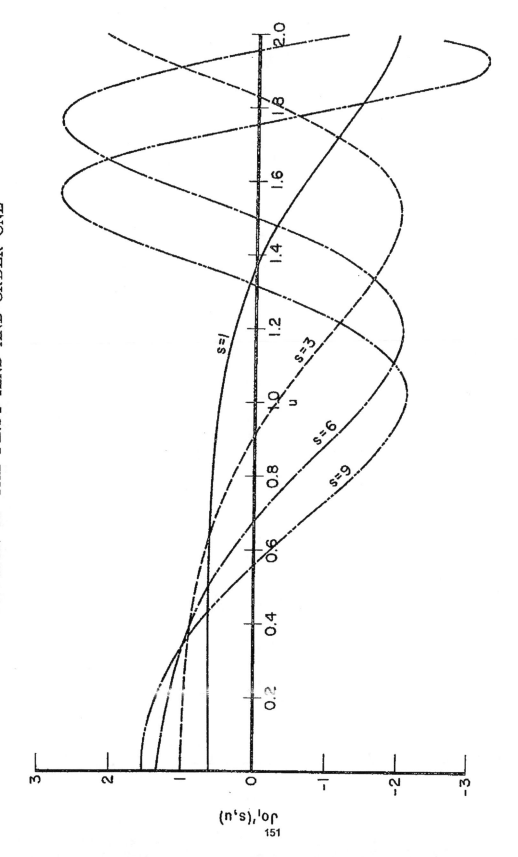

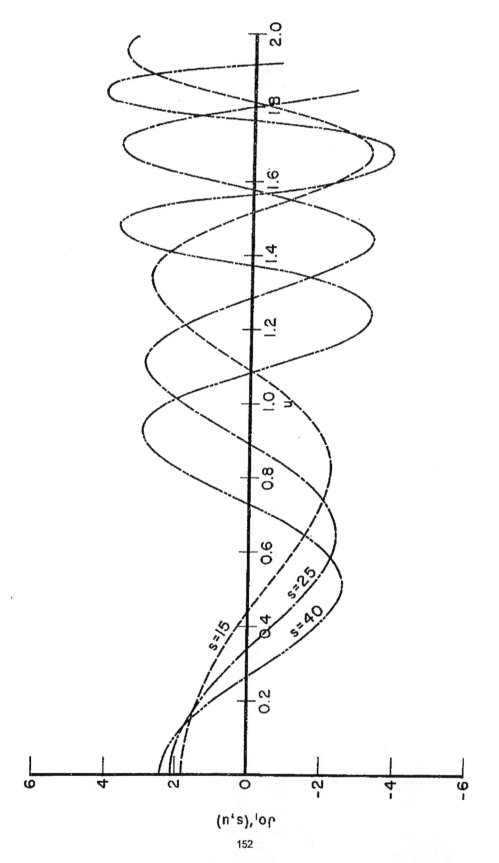

FIGURE 18    DERIVATIVE WITH RESPECT TO u OF THE "ODD" RADIAL MATHIEU
FUNCTION OF THE FIRST KIND AND ORDER ONE

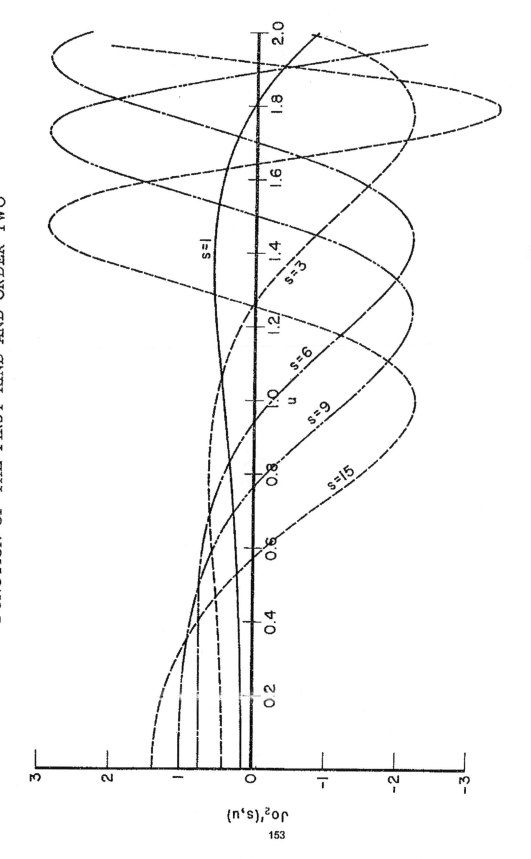

FIGURE 19    DERIVATIVE WITH RESPECT TO u OF THE "ODD" RADIAL MATHIEU
FUNCTION OF THE FIRST KIND AND ORDER TWO

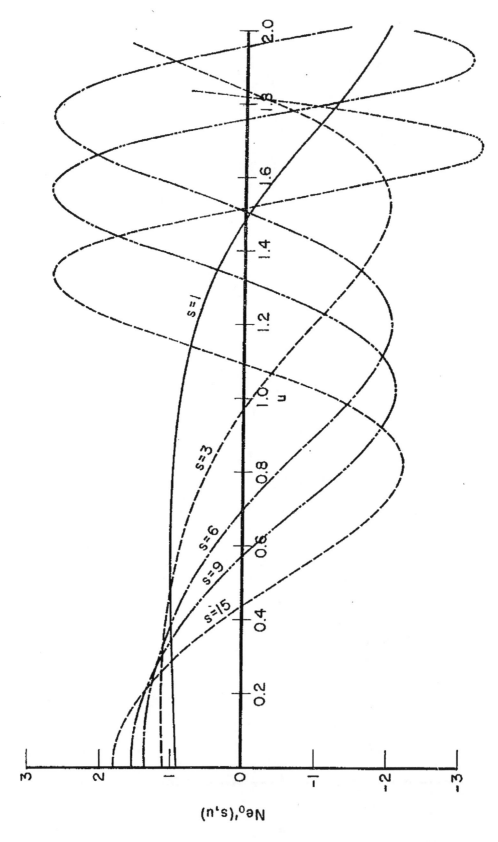

FIGURE 20    DERIVATIVE WITH RESPECT TO u OF THE "EVEN" RADIAL MATHIEU
FUNCTION OF THE SECOND KIND AND ORDER ZERO

FIGURE 21    DERIVATIVE WITH RESPECT TO u OF THE "EVEN" RADIAL MATHIEU
FUNCTION OF THE SECOND KIND AND ORDER ONE

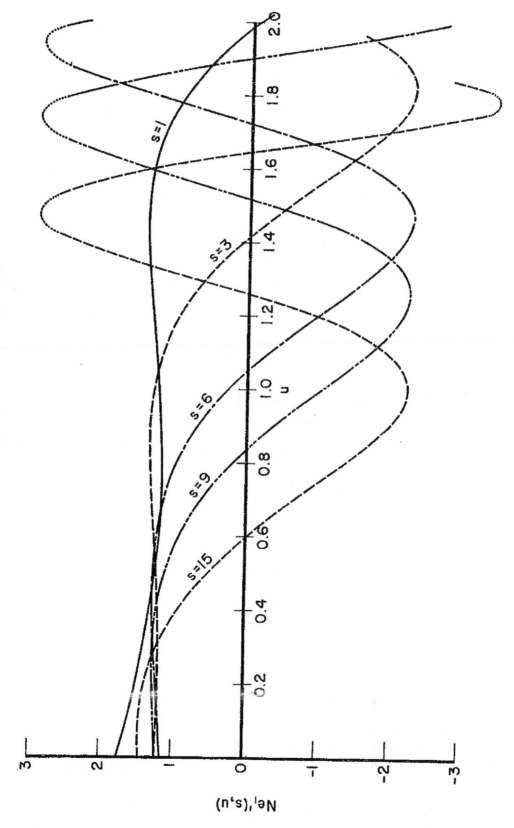

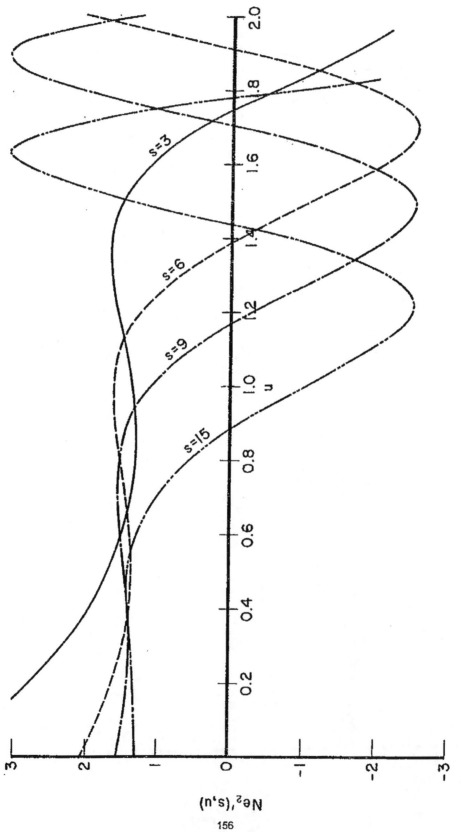

FIGURE 22    DERIVATIVE WITH RESPECT TO u OF THE "EVEN" RADIAL MATHIEU
FUNCTION OF THE SECOND KIND AND ORDER TWO

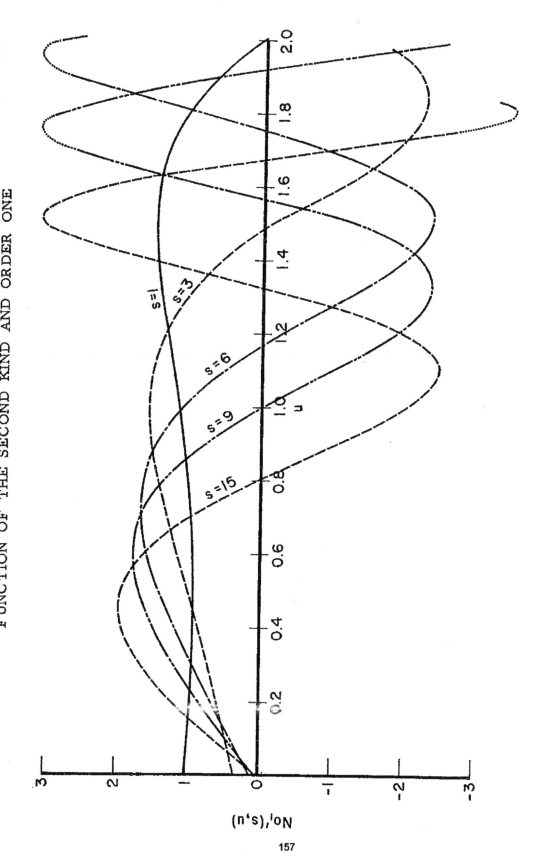

FIGURE 23    DERIVATIVE WITH RESPECT TO u OF THE "ODD" RADIAL MATHIEU
FUNCTION OF THE SECOND KIND AND ORDER ONE

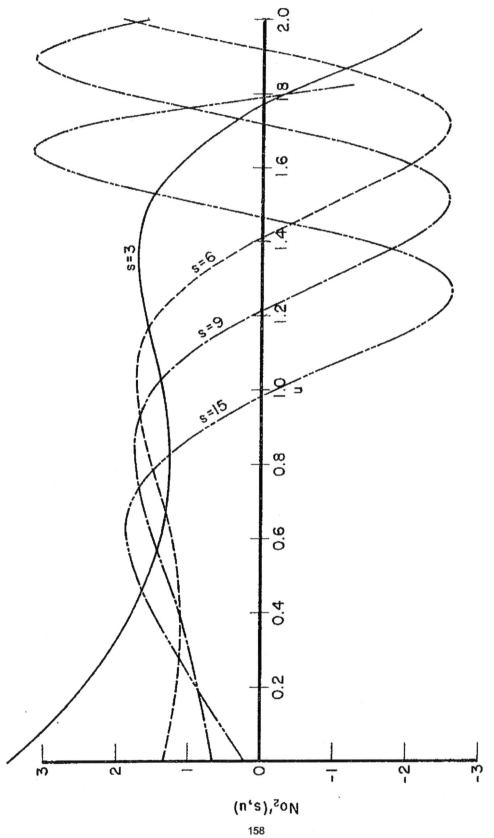

FIGURE 24    DERIVATIVE WITH RESPECT TO u OF THE "ODD" RADIAL MATHIEU
FUNCTION OF THE SECOND KIND AND ORDER TWO

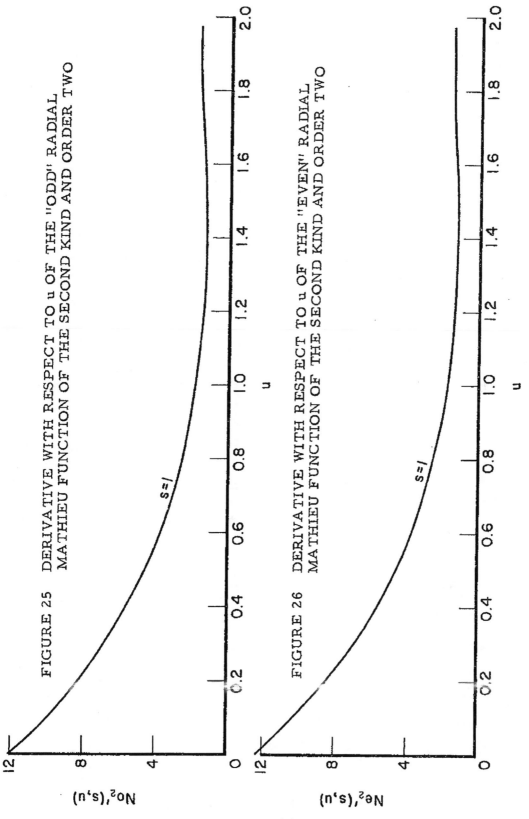

FIGURE 25     DERIVATIVE WITH RESPECT TO u OF THE "ODD" RADIAL MATHIEU FUNCTION OF THE SECOND KIND AND ORDER TWO

FIGURE 26     DERIVATIVE WITH RESPECT TO u OF THE "EVEN" RADIAL MATHIEU FUNCTION OF THE SECOND KIND AND ORDER TWO

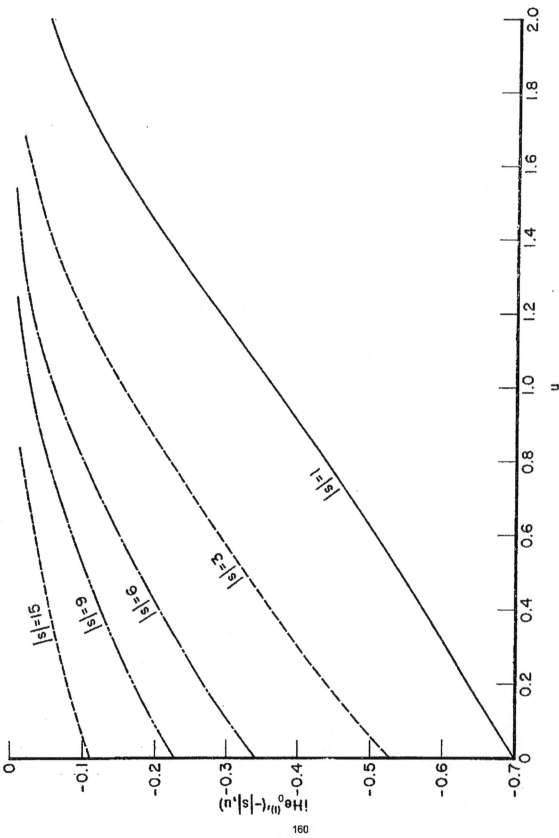

FIGURE 27    DERIVATIVE WITH RESPECT TO u OF THE "EVEN" RADIAL MATHIEU
FUNCTION OF THE THIRD KIND AND ORDER ZERO WITH NEGATIVE
PARAMETER·"s"

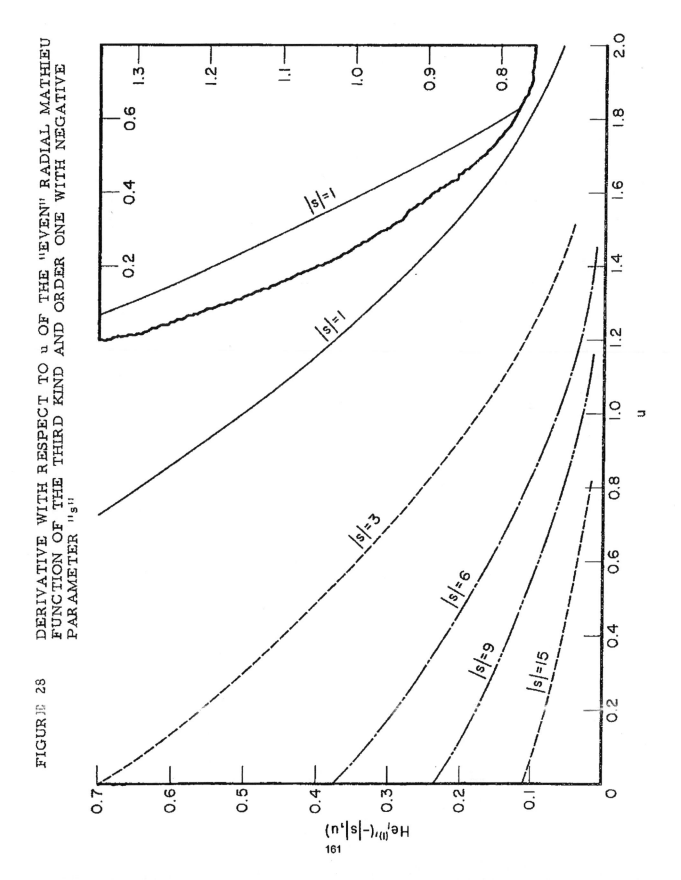

FIGURE 28    DERIVATIVE WITH RESPECT TO u OF THE "EVEN" RADIAL MATHIEU
FUNCTION OF THE THIRD KIND AND ORDER ONE WITH NEGATIVE
PARAMETER "s"

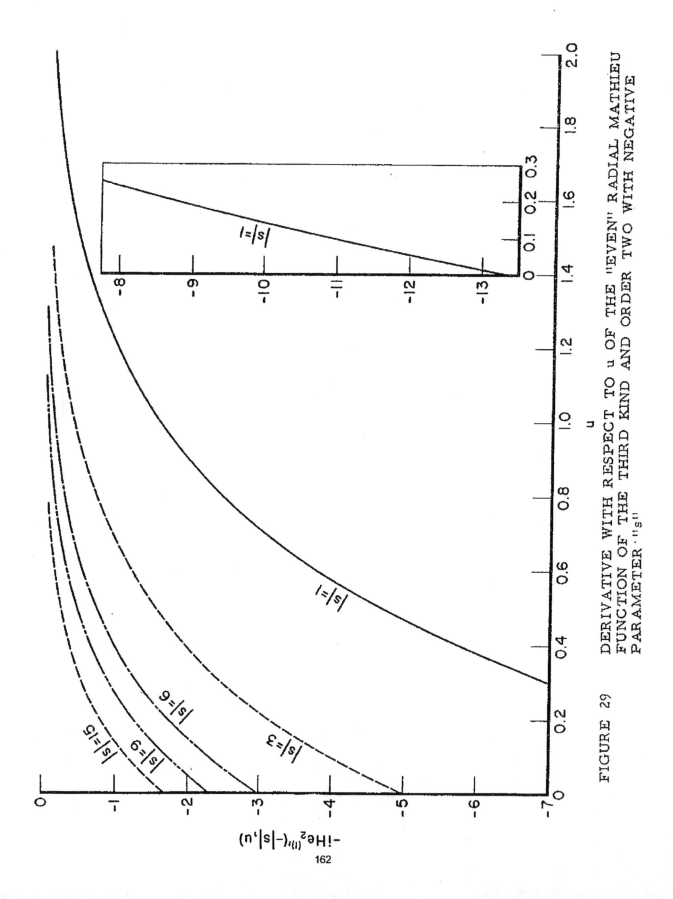

FIGURE 29    DERIVATIVE WITH RESPECT TO u OF THE "EVEN" RADIAL MATHIEU
FUNCTION OF THE THIRD KIND AND ORDER TWO WITH NEGATIVE
PARAMETER "s"

FIGURE 30    DERIVATIVE WITH RESPECT TO u OF THE "ODD" RADIAL MATHIEU
FUNCTION OF THE THIRD KIND AND ORDER ONE WITH NEGATIVE
PARAMETER "s"

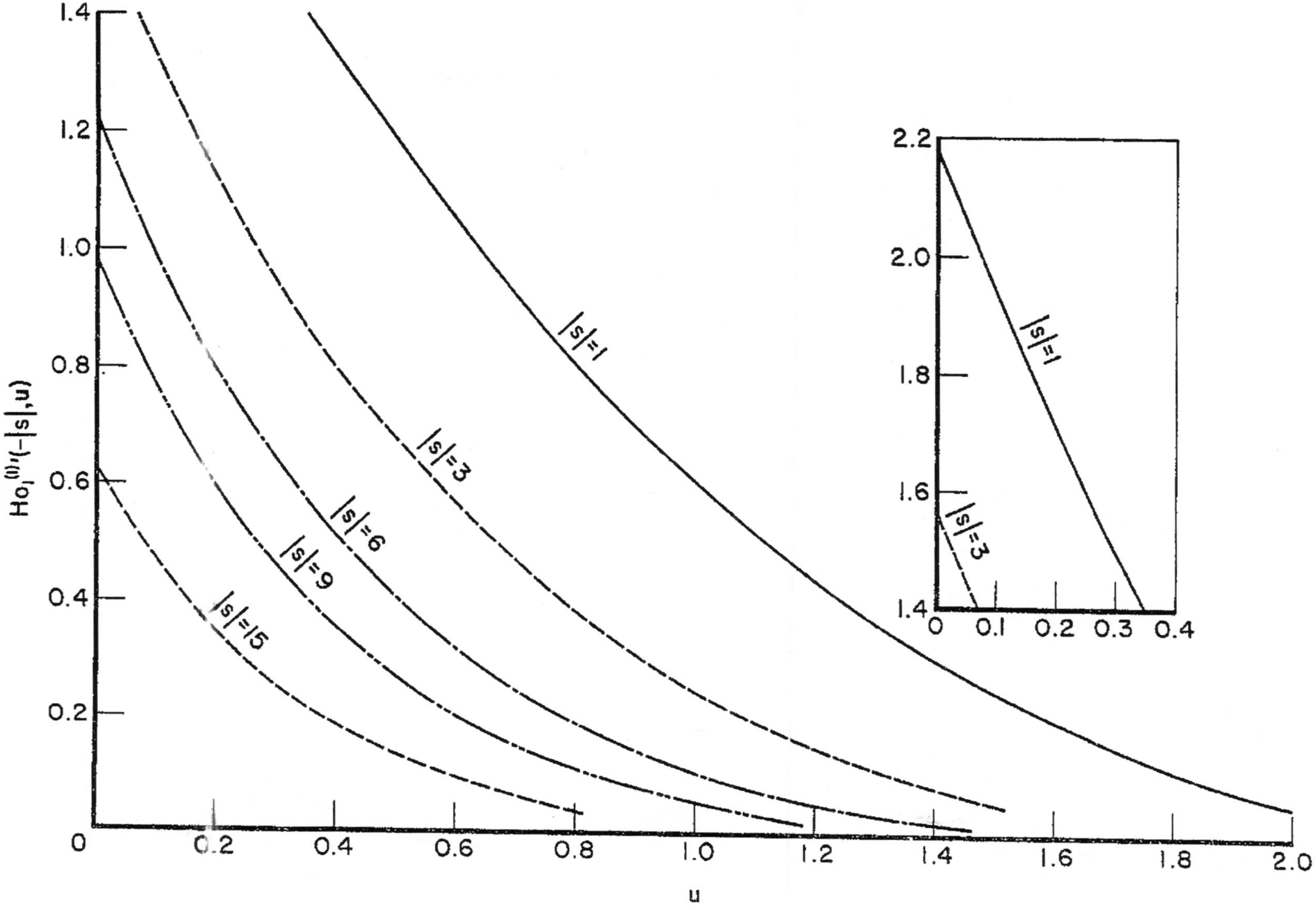

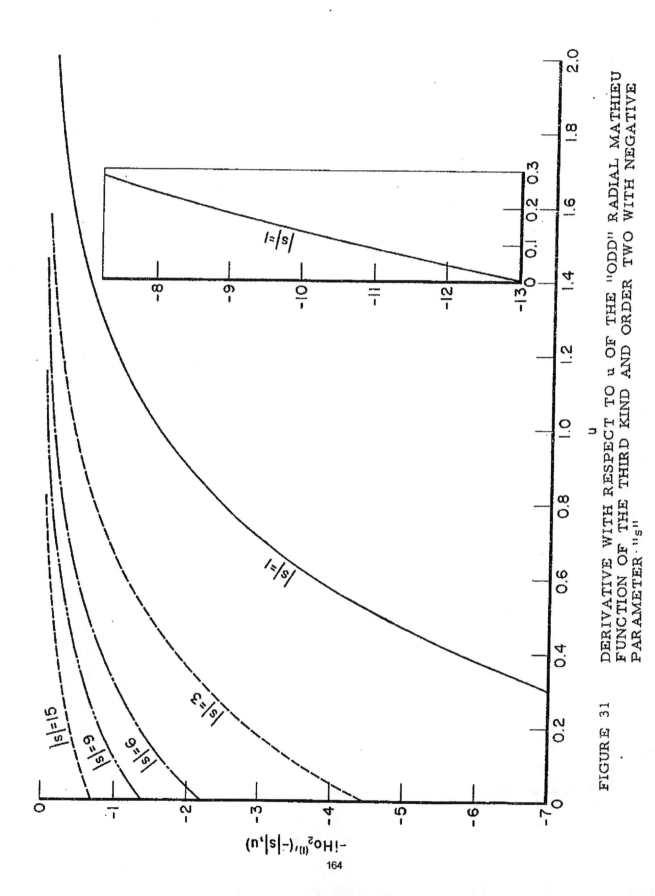

FIGURE 31    DERIVATIVE WITH RESPECT TO u OF THE "ODD" RADIAL MATHIEU FUNCTION OF THE THIRD KIND AND ORDER TWO WITH NEGATIVE PARAMETER "s"

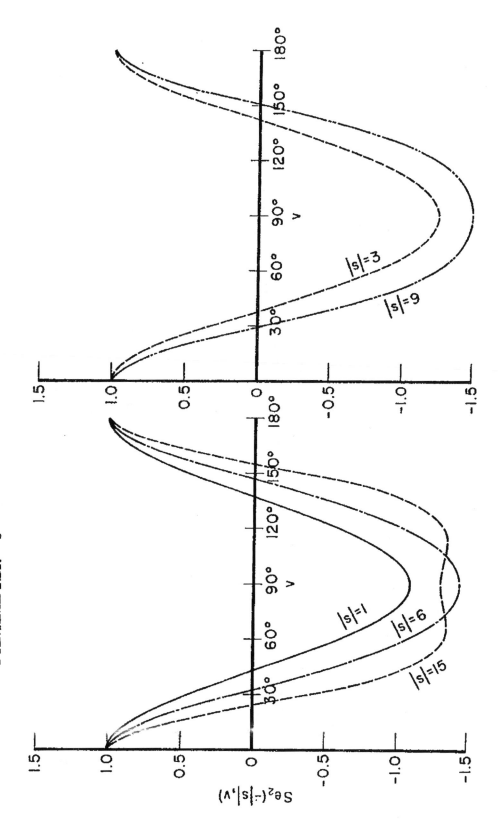

FIGURE 32    EVEN PERIODIC MATHIEU FUNCTION OF ORDER TWO WITH NEGATIVE
            PARAMETER "g"

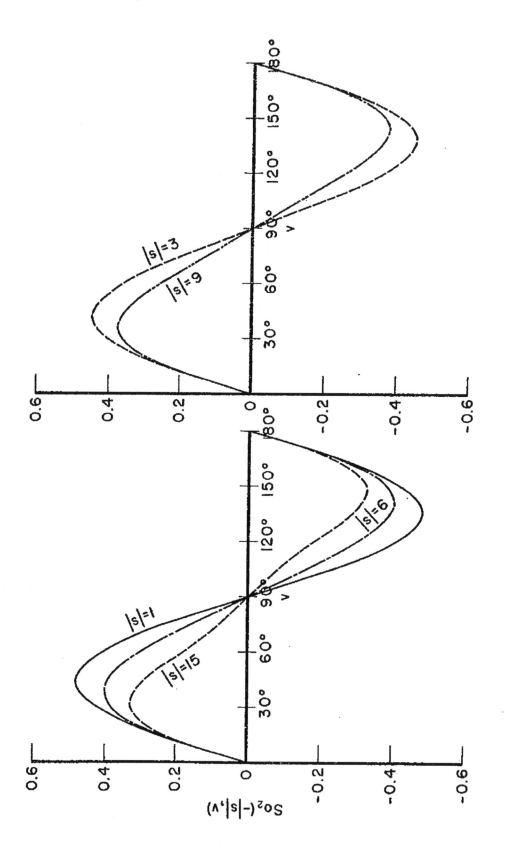

FIGURE 33    ODD PERIODIC MATHIEU FUNCTION OF ORDER TWO WITH NEGATIVE
PARAMETER "s"